AVALON 1997—THE CAMBRIAN STANDARD

THE UNIVERSITY OF THE STATE OF NEW YORK

Regents of The University

Avalon 1997—The Cambrian Standard

Third International Field Conference of the Cambrian Chronostratigraphy Working Group and I.G.C.P. Project 366 (Ecological Aspects of the Cambrian Radiation)

Edited by

Ed Landing
Geological Survey, New York State Museum
The State Education Department, Albany, New York 12230

and

Stephen R. Westrop
Department of Earth Sciences, Brock University
St. Catharines, Ontario L2S 3A1
Current address: Oklahoma Museum of Natural History and
School of Geology and Geophysics, University of Oklahoma
Norman, Oklahoma 73019

New York State Museum Bulletin 492
1998

The University of the State of New York
The State Education Department

Printed in the United States of America

Copies may be ordered from:
Publication Sales
New York State Museum
Albany, New York 12230
Phone: (518) 449-1404

Library of Congress Catalog Card Number: 97-62520

ISSN: 0278-3355
ISBN: 1-55557-209-X

This book is printed on acid-free paper.

Cover picture—Precambrian–Cambrian Global Stratotype Section and Point (GSSP) at Fortune Head, southeastern Newfoundland (see field trip stop on August 14).

Contents

Contents

Avalon 1997—A Pre-Meeting Viewpoint

Ed Landing
New York State Geological Survey, The State Education Department, Albany, NY 12230

Recent Developments In Cambrian Subdivision Philosophy

Following ratification of the Precambrian—Cambrian boundary Global Stratotype Section and Point (GSSP) in 1992 at the base of the *Trichophycus pedum* Zone at Fortune Head, eastern Newfoundland (Landing, 1994; formerly *Phycodes pedum*, see Geyer and Uchman, 1995), the focus of the International Subcommission on Cambrian Stratigraphy has been on defining globally useful chronostratigraphic divisions of the Cambrian. This effort has undergone several metamorphoses during the course of two earlier international field conferences on Cambrian subdivisions. The first field conference (**Morocco '95**) in May 1995 in southern Morocco was hosted by G. Geyer, E. Landing, and W. Heldmaier (Geyer and Landing, 1995), and was ostensibly conducted for the "Lower Cambrian Stage Subdivision Working Group". This concern with defining globally appropriate stage-level units was strongly promoted by A.Yu. Rozanov and supported by the previous officers of the Cambrian Subcommission.

However, a consensus developed on two themes during business meetings held during **Morocco '95**. The theme first was that participants in the "Lower Cambrian Stage Subdivision Working Group" had an interest in examining the upper limits of the Lower Cambrian and the transition into the Middle Cambrian. Because working groups on Middle and Upper Cambrian chronostratigraphy were not active, this led to an upward encroachment of the "Lower Cambrian Stage Subdivision Working Group" into the Middle Cambrian. The second theme that developed in Morocco featured two concerns: 1) whether or not the highly provincial marine faunas that developed on the numerous, widely separated continents of the Cambrian conveniently lend themselves to the development of a global stage-level chronostratigraphy, and, consequently, 2) if stage-level subdivisions were not best defined and limited to individual faunal provinces. Both of these themes had been discussed earlier (e.g., Landing, 1992).

This broadening of the "Working Group's" stratigraphic interest and a concern with defining provincial stages were evident in the second field conference, which was held in Spain (Linan et al., 1996) and also dealt with characteristic Gondwanan faunal and lithostratigraphic sequences. This conference featured a useful review of the Spanish Lower and Middle Cambrian stage-level nomenclature and its utility in Cadomian Europe for a more inclusive audience, now designated the "Cambrian Stage Subdivision Working Groups." Because of the relative abundance of biostratigraphically useful trilobites through the Middle Cambrian, the second conference had a strong Middle Cambrian focus.

Prospectus and Recommendations

Avalon 1997.—Gondwanan faunal and lithostratigraphic sequences are not the focus of the Cambrian Subcommission at the third field conference. The eastern Newfoundland and southern New Brunswick field areas of this conference lack any of the prominent Lower Cambrian carbonate-platform units and local evaporitic facies known in Spain and Morocco, and also lack any of the archaeocyath build-ups and any of the early trilobite groups that appear in the Lower Cambrian of Gondwana. For these reasons, the cool-marine faunas and lithostratigraphic sequences of eastern Newfoundland and southern New Brunswick are referable to a separate latest Proterozoic–Ordovician continent known as Avalon (Landing, 1996).

The Cambrian and Ordovician systems were first proposed in the Avalonian sequences of Wales and the Welsh Borderlands, and it is obviously incumbent to examine the faunal succession and geological history of the Avalon continent as part of an evaluation of Cambrian subdivisions. A long history of early paleontological research, as well as recent work throughout Avalonian North America and Britain by the organizers of **Avalon 1997**, demonstrate that southern New Brunswick and eastern Newfoundland have the most fossiliferous Cambrian sections in Avalon. These sections provide a basis for chronostratigraphic divisions of the Cambrian across Avalon and, potentially, for non-Avalonian regions.

Avalon 1997 will be more inclusive of the Cambrian than the first two working group field conferences, and field trip stops will illustrate strata as high as the Cambrian-Ordovician boundary and Tremadoc. For this reason, because working groups were never established for higher parts of the Cambrian, and because globally applied stage-level units may not be appropriate, a final metamorphosis of this working group into a "Cambrian Chronostratigraphy Working Group" seemed appropriate for this field conference.

Of particular significance for a global chronostratigraphy of the Cambrian is the regular appearance of datable volcanic ashes through Avalonian successions (Landing, 1994, 1996). We have recently begun to complete a series of reports on high-resolution U-Pb zircon dates from Avalonian Cambrian and Early Ordovician volcanic ashes (Isachsen et al., 1993; Landing et al., 1997, In press; Davidek et al., In press). A number of these dated ashes will be examined during this field trip. Our contention is that the limited resolution of biostratigraphic correlations demonstrates that the final arbiter of interregional Cambrian correlations will be high-resolution geochronology. The title of this field conference, "**Avalon 1997**–The Cambrian Standard", emphasizes that the presence of many volcanic ashes through fossiliferous uppermost Precambrian– Lower Ordovician sequences is unique to Avalonian North America and Britain, and that these regions will likely provide the best standard for interregional correlation of the Cambrian System.

Four Cambrian series and a Lower Cambrian Subsystem.—Avalon has not only provided the basis of the Cambrian and Ordovician Systems and the Precambrian—Cambrian boundary GSSP, but is also a reference standard for Cambrian subdivisions. The appropriate level, number, and definition of globally appropriate Cambrian subdivisions should be further discussed at **Avalon 1997**.

The organizers of **Avalon 1997** propose that faunal and geochronologic evidence show that the Cambrian of Avalon and other paleogeographic regions has four natural, series-level divisions. These series-level divisions of the Avalonian Cambrian include a traditional Acadian Series (Dawson, 1867, 1868) for the Middle Cambrian. "Acadian" was early applied to the paradoxidiid-bearing Cambrian strata of Avalonian Canada and the easternmost United States and has a base defined at or somewhat below the lowest occurrence of paradoxidiid trilobites with the appearance of characteristic Middle Cambrian trilobite groups. An appropriate series-level division for the Avalonian Upper Cambrian is a Merionethian Series, with a base defined near the lowest occurrence of *Agnostus pisiformis* (Cowie et al., 1972).

Recent geochronologic work indicates that the Lower Cambrian comprises the longest part of the Cambrian and is significantly longer than the Acadian and Merionethian combined. Two natural series-level divisions of the Avalonian Cambrian occur below the Acadian. These are the sub-trilobitic Placentian Series and the trilobite-bearing Branchian Series that were defined in eastern Newfoundland (Landing et al., 1989; Landing, 1992). The base of the Placentian is the base of the Cambrian, and has been defined at the Precambrian—Cambrian GSSP. The base of the Branchian Series lies at a regional unconformity overlain by the lowest occurrence of *Callavia broeggeri* Zone trilobites (Landing, 1992). These two series necessarily compose a higher level unit best regarded as a Lower Cambrian Subsystem. With a Cambrian base defined at approximately 543 Ma (Bowring et al., 1993) and a Lower–Middle Cambrian boundary younger than 511 Ma (Landing et al., In press), this Lower Cambrian Subsystem is significantly longer than 30 Ma. Indeed, a recent U-Pb zircon date of 490.5 ± 1.5

Ma on the upper Upper Cambrian, middle *Peltura scarabaeoides* Zone (Davidek et al., In press) supports Shergold's (1995) proposal that the Cambrian–Ordovician boundary is younger than 490 Ma, and assigns only about 20 Ma to the Middle and Late Cambrian.

Acknowledgments

Preparation and organization of **Avalon 1997** were done during the tenure of National Science Foundation grant EAR94-15773 (to E. Landing) and National Sciences and Engineering Research Council of Canada grant 41197 (to S.R. Westrop).

References

Bowring, S.A., J.P. Grotzinger, C.E. Isachsen, A.H. Knoll, S.M. Pelechaty, and P. Kolosov. 1993. Calibrating rates of Early Cambrian evolution. Science, 261:1293–1298.

Cowie, J.W., A.W.A. Rushton, and C.J. Stubblefield. 1972. A correlation of the Cambrian rocks in the British Isles. Geological Society Special Report No. 2, 42 p.

Davidek, K.L., E. Landing, S.A. Bowring, S.R. Westrop, A.W.A. Rushton, R.A. Fortey, and J. Adrain. In press. New uppermost Cambrian U-Pb date from Avalonian Wales and age of the Cambrian–Ordovician boundary. Geological Magazine.

Dawson, J.W. 1867. On recent geological discoveries in the Acadian provinces of British America. Proceedings of the American Association for the Advancement of Science, 16:118 (issued in 1868).

————. 1868. Acadian Geology. MacMillan and Company, London, 684 p.

Geyer, G., and E. Landing (eds.). 1995. Morocco '95—The Lower–Middle Cambrian standard of western Gondwana. Introduction, field guide, abstracts, and proceedings of the First Conference of the Lower Cambrian Stage Subdivision Working Group and I.G.C.P. Project 366 (Ecological Aspects of the Cambrian Radiation). Beringeria Special Issue 2, 269 p.

————, and A. Uchman. 1995. Ichnofossil assemblages from the Nama Group (Neoproterozoic–Lower Cambrian) in Namibia and the Proterozoic–Cambrian boundary problem revisited, p. 175–202. *In* G. Geyer and E. Landing (eds.), Morocco '95—The Lower–Middle Cambrian Standard of Western Gondwana. Beringeria Special Issue 2, 269 p.

Isachsen, C.E., S.A. Bowring, E. Landing, and S.D. Samson. 1994. New constraint on the division of Cambrian time. Geology, 22:496–498.

Landing, E. 1992. Lower Cambrian of eastern Newfoundland: epeirogeny and Lazarus faunas, lithofacies-biofacies linkages, and the myth of a global chronostratigraphy, p. 283—309. *In* J.H. Lipps and P.W. Signor (eds.), Origin and Early Evolution of the Metazoa. Plenum Press, New York.

————. 1994. Precambrian—Cambrian boundary ratified and a new perspective of Cambrian time. Geology, 22:179–182.

————. 1996. Avalon: Insular continent by the latest Precambrian, p. 29–63. In R.D. Nance and M.D. Thompson (eds.), Avalonian and Related Perigondwanan Terranes of the Circum-North Atlantic. Geological Society of America Special Paper 304.

————, P. Myrow, A.P. Benus, and M.M. Anderson. 1989. The Placentian Series: appearance of the oldest skeletalized faunas in southeastern Newfoundland. Journal of Paleontology, 63:739–769.

————, S.A. Bowring, R.A. Fortey, and K.L. Davidek. 1997. U-Pb zircon date from Avalonian Cape Breton Island and geochronologic calibration of the Early Ordovician. Canadian Journal of Earth Scienecs, 34:724—730.

————, ————, K. Davidek, S.R. Westrop, G. Geyer, and W. Heldmaier. In press. New U-Pb zircon dates from Avalon and Gondwana and duration of the Early Cambrian. Canadian Journal of Earth Sciences.

Linan, E., J.A. Gamez Vintaned, and R. Gozalo. 1996. II Field Conference of the Cambrian Stage Subdivision Working Groups. International Subcommission on Cambrian Stratigraphy. Spain, 13–21 September 1996. Field trip guide and abstracts. University of Zaragoza, Zaragoza, 129 p.

Shergold, J. 1995. Timescales. Calibration and developments, 1. Cambrian. Australian Geological Survey Organization, Record 1995/30, 36 p.

MEETING ACKNOWLEDGMENTS

Avalon 1997 was held during a time when funding for basic research on topics as fundamental as the history and evolution of life continued to decrease. For this reason, the organizers of **Avalon 1997** extend our deep appreciation for the financial and technical aid from a number of organizations and individuals that supported this international field conference.

FINANCIAL SUPPORTERS INCLUDED:

Geological Survey of Canada (via Drs. James M. Franklin, Chief Scientist, and Godfrey S. Nowlan, Research Scientist)

International Union of Geological Sciences (via Zdenek Johan, Treasurer)

New Brunswick Museum (via Drs. Frank Milligan, Director, and Randall F. Miller, Curator of Geology and Paleontology)

New York State Museum (via Dr. Louis D. Levine, Director).

VEHICLES WERE PROVIDED FOR THE FIELD EXCURSIONS BY:

New Brunswick Geological Surveys (via Dr. Les Ffyfe, Manager, Geological Surveys [South])

Government of Newfoundland and Labrador, Department of Mines and Energy (via Paul L. Dean, Assistant Deputy Minister, Mines; D. Wayne Ryder, Director, Project Management; Sidney J. Parsons, Administrator (Newfoundland), Project Management; Karen Dawson, Accounting Clerk, Project Management; and W. Douglas Boyce, Provincial Paleontologist).

THANKS ARE EXTENDED TO THE FOLLOWING:

Michael Lozon, Department of Earth Sciences, Brock University, St. Catharines, Ontario, efficiently converted rough drafts of the figures of this bulletin into publication-ready copy.

Jeanne C. Finley, Linda Van Aller Hernick, Donna Jornov, and John B. Skiba of the New York State Geological Survey assisted in preparation of the draft copy of New York State Museum Bulletin 492 for the field conference.

W. Douglas Boyce, Government of Newfoundland, Department of Mines and Energy, and Randall F. Miller, New Brunswick Museum, generously helped with meeting coordination.

Dr. Bruce Runnegar, Department of Geosciences, University of California, Los Angeles, provided a public lecture on the significance of Ediacaran faunas for the New Brunswick Museum in the course of the conference.

Partial page charges for publication of New York State Museum Bulletin 492 were received from National Science Foundation grant EAR94-15773 (Principal Investigator, Ed Landing).

G. Geyer and A.R. Palmer provided peer reviews of this bulletin.

Cambrian Faunal Sequence and Depositional History of Avalonian Newfoundland and New Brunswick: Field Workshop

Ed Landing and Stephen R. Westrop[*]
New York State Geological Survey, The State Education Department,Albany, NY 12230, and
[*]*Department of Earth Sciences, Brock University,St. Catharines, Ontario L2S 3A1*
Current Address: Oklahoma Museum of Natural History and School of Geology and Geophysics, University of Oklahoma, Norman, Oklahoma 73019

Depositional Setting of Avalonian Cambrian

Avalon, insular latest Precambrian–Ordovician continent.—The Avalon terrane of the coastal northeastern United States, Maritime Canada, eastern Newfoundland, and southern Britain was originally an insular latest Precambrian–Ordovician continent that collided with the eastern margin of Laurentia during the Middle Paleozoic Acadian-Caledonian orogeny. Three features led to the original recognition of Avalon as an exotic terrane on the southeastern margin of the Caledonian-Acadian orogen: 1) a basement dominated by thick late Proterozoic volcanic and siliciclastic sequences, 2) evidence for later Proterozoic compression and intrusion during an "Avalonian orogeny," and 3) a post-Avalonian orogeny cover sequence of siliciclastic-dominated Cambrian–Silurian rocks with provincially distinctive faunas (Rast et al., 1976).

Avalon was one of a number of continents that originated during the interval of latest Precambrian rifting and drift. Latest Precambrian—Ordovician outliers in the coastal northeastern Appalachians (Figure 1), Wales, and southern England show geographically widespread similarities in this cover sequence (e.g., ten epeirogenic sequences, coeval volcanic episodes; lithosomes that blanket large parts of Avalon and allow the recognition of identical group-, formation-, and member-level units from Rhode Island to southern Britain; and characteristic cool-water Avalonian province faunas) that demonstrate that Avalon was a unified continent-like unit. Although timing of the Avalonian orogeny was similar to that of the Pan-African and Cadomian orogenies, the lithostratigraphy of its siliciclastic-dominated, shallow-marine, uppermost Proterozoic–Lower Paleozoic cover sequence and distinctiveness of its faunas indicate that Avalon had no relationship to Gondwana (inclusive of Cadomian Europe, which shows prominent Lower Cambrian carbonate platform deposits with archaeocyaths, a group unknown in Avalon) or to Baltoscandia (which has no evidence of an Avalonian orogeny or any of the characteristic lithofacies of the Avalonian Lower Cambrian) (see Landing, 1996a). Although Avalonian Cambrian faunas are typically assigned to an "Atlantic" or "Acado-Baltic" faunal province (Palmer, 1969; Rast et al., 1976), Cambrian and Ordovician faunal distinctiveness between the Avalon and Baltica continents (Conway Morris and Rushton, 1988, p. 98) was defined by faunal barriers, presumably related to "their separation by a deep ocean, unfavorable ocean currents, or some other barrier". Theokritoff (1979) was correct to distinguish an Avalonian from a Scandian faunal province and to abandon the term "Acado-Baltic."

Epeirogenic activity and Cambrian deposition.—Deposition of the Avalonian uppermost Precambrian–Lower Paleozoic seems to have

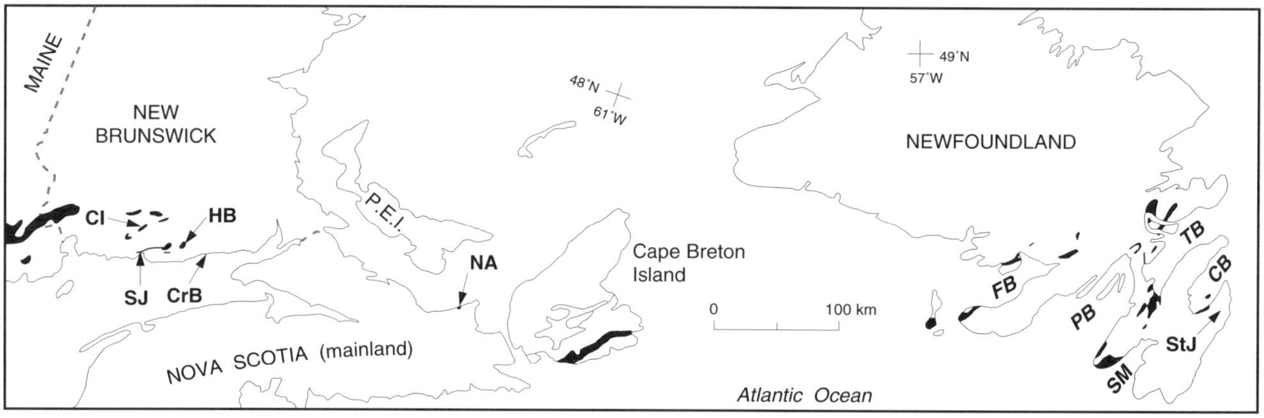

Figure 1. Uppermost Proterozoic–Lower Paleozoic cover sequence (in black) in Avalonian Maine, Maritime Canada, and southeast Newfoundland. *CB,* Conception Bay; CI, Caton's Island; CrB, Cradle Brook; *FB,* Fortune Bay; HB, Hanford Brook; NA, northern Antigonish Highlands; *PB,* Placentia Bay; P.E.I., Prince Edward Island; SJ, Saint John; *SM,* StJ, St. John's; *SM,* Saint Mary's Bay; *TB,* Trinity Bay. Modified from Landing (1996b, fig. 1).

been in an active transtensional regime that: 1) controlled the distribution, thickness, and local stratigraphic continuity of the cover sequence; 2) determined the location and timing of minor mixed (basaltic and rhyolitic) volcanism; and 3) largely obscured the influence of eustatic events by movement on growth faults (Landing and Benus, 1988b; Landing, 1996a). Secular changes in the type of Cambrian volcanism from latest Proterozoic–Early Cambrian rhyolites, to middle Middle Cambrian basalts, and back to rhyolite volcanism in the Late Cambrian (e.g., Landing 1996a, Landing et al., In press) and the transtensional regime of Avalon suggest a back-arc setting adjacent to a strike-slip fault, a setting similar to but much shallower than the Neogene Cretan Sea basin (see Pe-Piper et al., 1995).

The oldest cover sequence deposits are limited to northern and western outliers and define the "marginal platform" of Avalon (Landing, 1996a). These northerly and westerly outliers in southern New Brunswick (Figure 1, localities CI, SJ, HB), eastern Newfoundland (Figure 1, southern Burin Peninsula and Fortune Bay [FB]), and the "Welsh geosyncline" of North Wales have thick uppermost Precambrian through lowest Cambrian sequences that were deposited in down-dropped, fault-defined basins with the earliest extension of the Avalonian orogen (e.g., Smith and Hiscott, 1984).

Latest Proterozoic–early Placentian deposition.—Stratigraphic continuity exists through fossiliferous Precambrian-Cambrian boundary interval strata only in marginal platform successions, which include up to 2.5 km of lower fault-basin red beds and marginal marine facies (Rencontre Formation) and up to 1.0 km of overlying storm-dominated shelf deposits of the Chapel Island Formation in southeastern Newfoundland (see Myrow et al., 1988). A somewhat thinner Rencontre–Chapel Island succession is present in southern New Brunswick (Figure 1, localities HB and SJ; see August 18, Somerset Street, and August 19, Hanford Brook East). The Precambrian–Cambrian global stratotype section and point (GSSP) (August 14, Fortune Head) was defined in one of these marginal platform successions in the fault-bounded Fortune Bay–Burin Basin at Fortune Head, eastern Newfoundland (Figure 2, locality FH; August 14, Fortune Head; see Landing, 1994).

The uppermost Proterozoic–lower Placentian cover sequence comprises two depositional sequences in Avalonian North America. In southeastern Newfoundland and southern New Brunswick, shoaling of the Chapel Island shelf was followed by erosion and subsequent deposition of the initial beds of sequence 2, either as a red limestone on a soil horizon developed on red siliciclastic mudstone (August 14, Little Dantzig

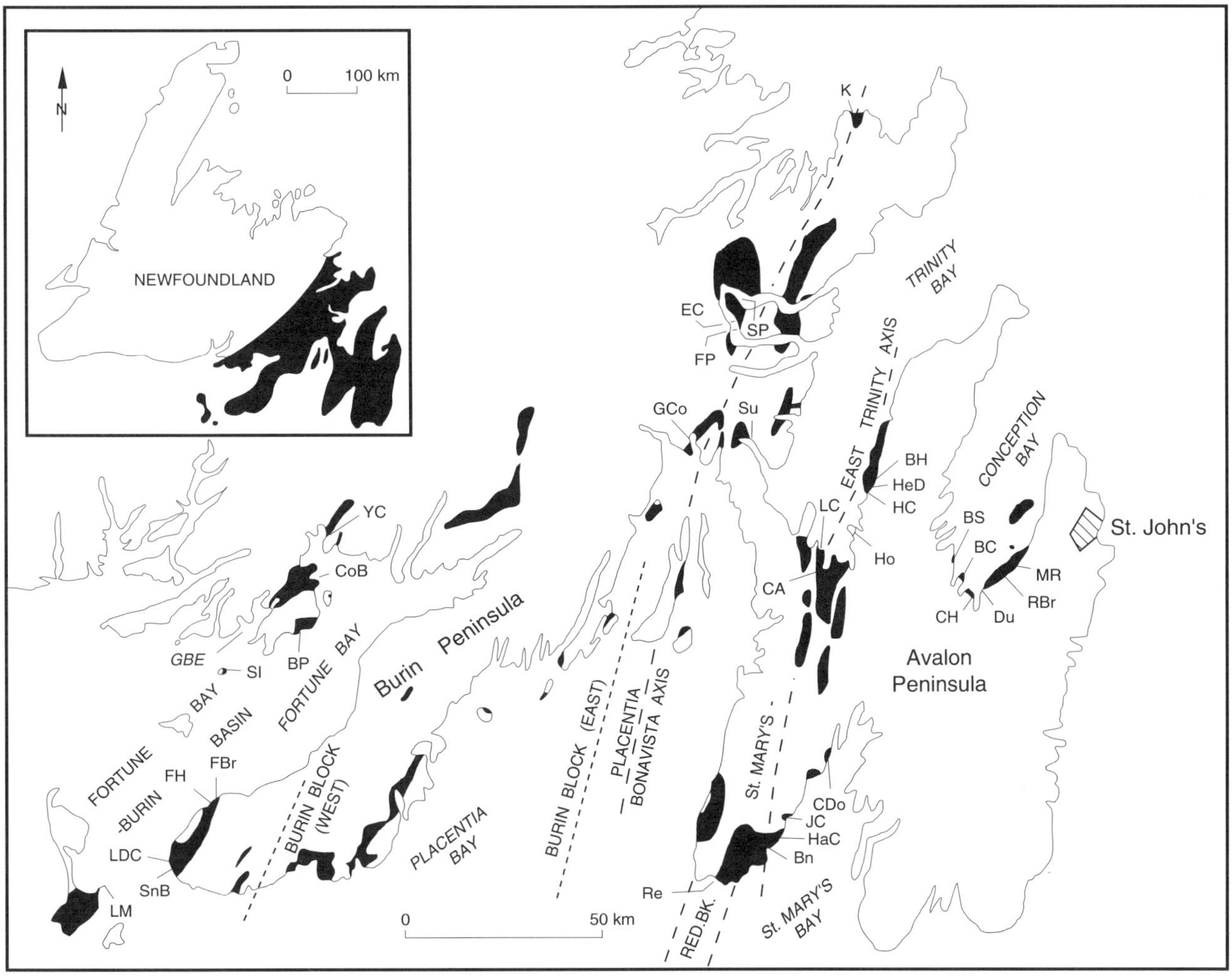

Figure 2. Uppermost Precambrian–Lower Ordovician (black on large scale map) and structural units in southeast Newfoundland. Abbreviations: BC, Bacon Cove; BH, Boar's Head; Bn, Branch; BP, Blue Pinion Cove; BS, Brigus Point South; CA, Chapel's Arm; CDo, Cape Dog; CH, Chapel's Head; CoB, Corbin Bay; Du, Duffs; EC, Elliott Cove; FBr, Fortune Brook; FH, Fortune Head; FP, Fosters Point; *GBE*, Grand Bay de l'Eau; GCo, Goose Cove; HaC, Hay Cove; HC, Highland Cove; HeD, Heart's Delight; Ho, Hopeall Head; JC, Jigging Cove; K, Keels; LC, Long Cove; LDC, Little Dantzig Cove and Little Dantzig Cove Brook; LM, Langlade; MPE, McLeod Point; MR, Manuels River; RBr, Red Brook Road; Re, Redlands Cove; RED.BK., Redlands Block; SI, Sagona Island; SnB, Snook's Brook; SP, Smith Point; Su, Sunnyside; YC, Young's Cove. Figure modified from Landing (1996a, fig. 1).

Cove) or as conglomeratic, stream-channel fills (August 19, Hanford Brook East). The lower deposits of sequence 2 represent a return to feldspathic, wave-dominated sandstones in Avalonian North America. An abrupt switch from wave-dominated to macrotidal environments and an influx of relatively coarse-grained quartz sand mark the later part of depositional sequence 2, and the accumulation of the Random Formation across Avalonian North America (Figure 3) and coeval, lithologically comparable units in Avalonian Britain (Landing, 1996a).

Cambrian depocenters shifted progressively eastward in Avalonian North America and Britain during the Cambrian after accumulation of depositional sequence 2. This shift indicates that transtensional activity, and not simple rifting along one locus (see Steel and Gloppen, 1980),

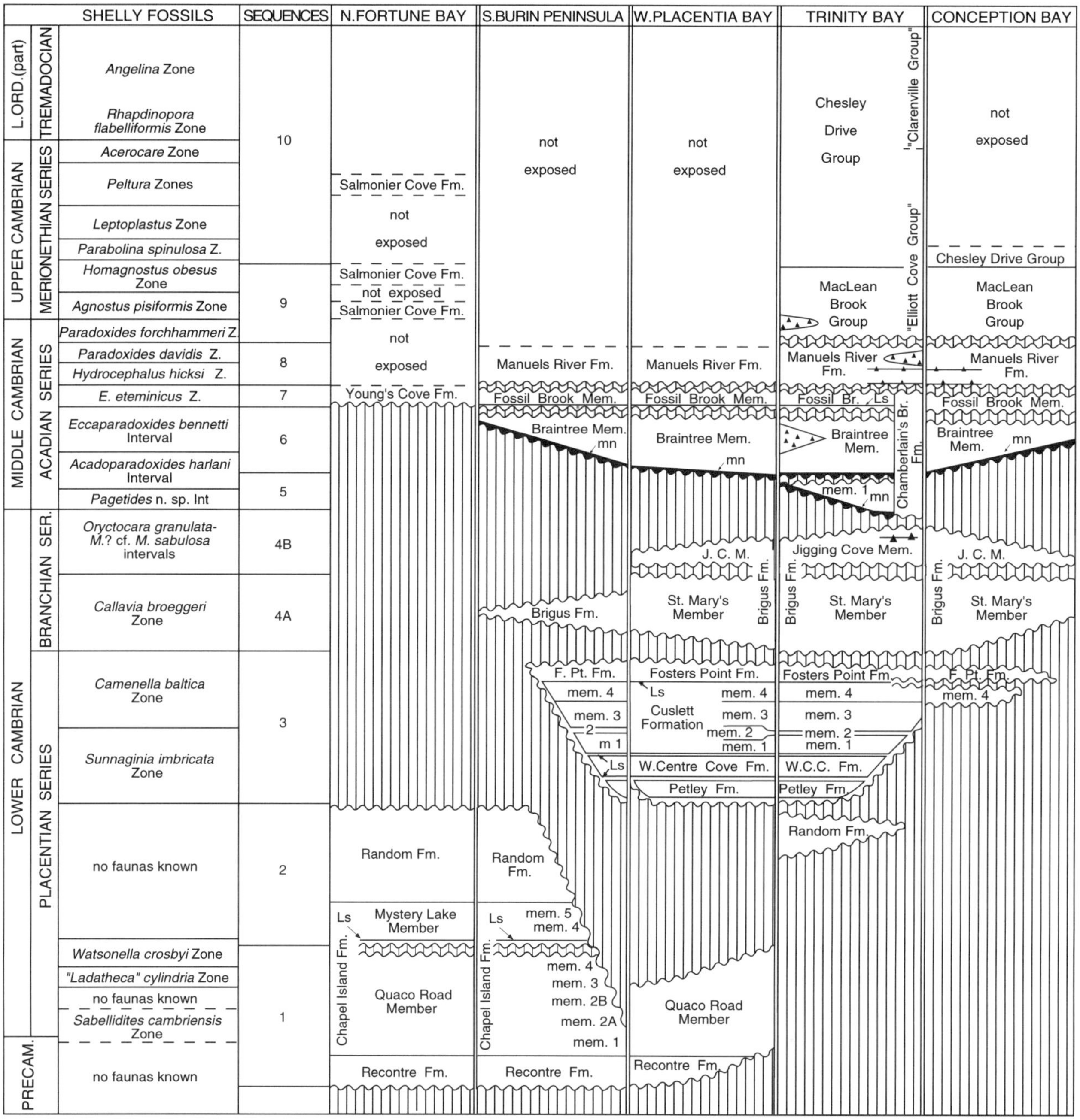

controlled Lower Paleozoic deposition. Post-Random basin reorganization initially led to uplift on the Burin and Redlands blocks (Figure 2) and local erosional truncation of the Random Formation (August 17, Sunnyside). This erosional truncation defines an important epeirogenic sequence boundary (depositional sequence 2–3 unconformity) across Avalon (Landing, 1996a).

Late Placentian deposition.—Subsequent extension led to development of the narrow, roughly N–S-trending Placentia–Bonavista axis almost 100 km east of the Fortune Bay–Burin Basin in eastern Newfoundland (Figure 2). Deposition of up to several hundred meters of the shallow-marine, siliciclastic mudstone-dominated Bonavista Group (depositional sequence 3) took place in this basin on the inner platform of Avalon during the late Placentian Epoch. Inner platform successions have strata no older than the Random Formation as the oldest cover sequence unit, and include the Placentia–Bonavista axis and entire Avalon Peninsula in eastern Newfoundland (Landing and Benus, 1988b).

Inner platform successions are recognized at the easternmost limit of the Cambrian in southern New Brunswick (Landing, 1996b). In addition, the Shropshire and Warwickshire Cambrian outliers, where the Random Formation (known locally as Wrekin Quartzite and lower Hartshill Formation [Park Hill, Tuttle Hill, and Jees Members]; e.g., Brasier, 1984, 1989) rests on the Avalonian orogen as the oldest Cambrian unit, are also referable to the inner platform.

Progressive onlap of the upper Placentian east and west of the Placentia–Bonavista axis took place after definition of this depocenter. This onlap seems to have been driven by continued strike-slip activity and subsidence (Landing and Benus, 1988a). Outcrops that will be visited on this trip show fissure-fills of upper Placentian sediments into the Random Formation (August 17, Sunnyside) and late Precambrian Holyrood granite (August 13, Duffs and Chapel Cove) that provide evidence of this extension and demonstrate the greater duration of the sub-Bonavista Group hiatus away from the Placentia—Bonavista axis (Figure 3).

A feature characteristic of the Lower Cambrian of Avalon is the frequent deposition of a limestone as a initial onlap deposit. These limestones are invariably thin (i.e., rarely more than a meter thick), condensed, amalgamated beds; carbonate platform sequence lithologies (e.g., massively bedded limestones, dolostones, evaporitic facies) are not part of any Avalonian Cambrian–Lower Ordovician sequence. At all of the stops in this field conference, limestones (except methanogenic carbonate nodules) represent the shallowest facies in Avalonian shale basins, and are found as on-shore deposits, the tops of shoaling-up cycles, or the initial deposit of a transgressive systems tract (e.g., Landing and Benus, 1988b; Landing et al., 1989; Myrow and Landing, 1992). Shell-hash limestones in the Lower Cambrian occur as interbeds in red or purple-red siliciclastic mudstones. This is significant because Avalonian Cambrian shales exhibit a spectrum of colors that includes the series: black–dark to light grey–green–purple–red. Except when affected by hydrothermal activity adjacent to faults or shears, this color spectrum accurately records progressively shallower water and, indirectly, relative distance from shore (Landing et al., 1989; Myrow and Landing, 1992). Field trip participants will often be able to gauge relative water depths in the course of this conference by noting shale color.

Branchian deposition.—The final eastern

←
Figure 3. Epeirogenic sequence stratigraphy in southeast Newfoundland. Fossil zonation from Landing (1992); Middle Cambrian–Tremadocian zones modified from Hutchinson (1952, 1962), Fletcher (1972), and Rushton (1974). Precambrian–Cambrian global stratotype defined in lower *Sabellidites cambriensis* Zone (body fossils) at base of *Trichophycus pedum* Zone (formerly *Phycodes pedum* Zone; see Geyer and Uchman, 1995) at Fortune Head, southern Burin Peninsula (Narbonne et al., 1987; Landing, 1994). Bonavista Group includes Petley, West Centre Cove, Cuslett, and Fosters Point Formations (Landing and Benus, 1988b). See Landing (1996a) for use of Braintree and Fossil Brook Members of the Chamberlain's Brook Formation and replacement of "Elliott Cove" and "Clarenville" Groups by MacLean Brook Group (this volume) and Chesley Drive Group. Manganese nodule bed at depositional sequence 6 base is Fletcher's (1972) "Easter Cove Blister bed"; limestone (Ls) base of sequence 7 is Fletcher's (1972) "Deep Cove Limestone". Vertical lines show hiatus; wavy line is unconformity; black triangles on lines are ashes; scattered black triangles are volcanic lenses (=Cape Dog and Hopeall Head volcanics in upper Chamberlain's Brook Formation, Hay Cove volcanics in upper Manuels River Formation, Chapel Arm Member in MacLean Brook Group. Abbreviations: F.Pt., Fosters Point; J.C.M., Jigging Cove Member; Ls, limestone; mn, manganese nodule bed; W.C.C., West Centre Cove. Modified from Landing (1996a, fig. 2).

shift of depocenters on the inner platform took place during the latest Placentian and earliest part of the trilobitic late Early Cambrian (Branchian Epoch) of Avalon, and controlled deposition through the remainder of the Cambrian and Ordovician. A further 50 km eastern shift of depocenters occurred with definition of the West St. Mary's—east Trinity axis (Figure 2), which has the thickest deposits of the Branchian Series and Middle Cambrian (Acadian Series) (Hutchinson, 1962; Landing, 1996a). The relatively thick Branchian Series—lowest Ordovician in Warwickshire (Taylor and Rushton, 1971; Rushton, 1974) closely resembles the succession in the east St. Mary's-west Trinity axis, and the "Warwickshire trough" (Brasier, 1989) may have been continuous with the St. Mary's–east Trinity axis. Landing and Benus (1988b) noted that the thickest deposits of the terminal Placentian Series unit, or Fosters Point Formation in Avalonian North America, lie in eastern Trinity Bay, and definition of the west St. Mary's–east Trinity axis may have taken place in the latest Placentian and had a role in producing the regional sequence boundary between the Placentian and Branchian Series (Figure 3).

Onlap of the Branchian Series took place east and west of the west St. Mary's–east Trinity axis. Only relatively thin Branchian and Acadian series units reached the marginal platform and unconformably overlie units as low as the middle Placentian Series Random Formation in the Burin Peninsula (August 14, Little Dantzig Cove Brook) or Saint John, New Brunswick area (August 18, Somerset Street; August 19, Hanford Brook).

Although earlier interpreted as a single, shale-dominated depositional sequence (number 4) (Landing, 1996a), the upper Lower Cambrian Branchian Series is now regarded as comprising two sequences (Figure 3; August 15, Jigging Cove, and August 16, Fosters Point). In eastern Newfoundland, the lower depositional sequence (or St. Mary's Member [new]) is the *Callavia broeggeri* Zone. The abrupt faunal change from the *Callavia broeggeri* Zone into the overlying *Myopsostrenua?* cf. *M.? sabulosa* Faunal Interval is a sequence boundary marked by a phosphate pebble bed, sandstone, or debris flow at the base of the upper member (Jigging Cove Member [new]) of the Brigus Formation (see Landing and Westrop, this volume).

Post-Branchian deposition.—Post-Early Cambrian deposits on the inner and outer platforms are initially limited to shales. The lower Acadian Series on the inner platform includes a lower green and red shale-dominated Chamberlain's Brook Formation, characteristically with manganese carbonate and oxide nodules through most of its lower and middle parts, and which extends from Rhode Island to eastern Newfoundland in Avalonian North America. An unconformity of variable magnitude exists at all contacts of the Chamberlain's Brook Formation with upper Lower Cambrian depositional sequence 4, and stratigraphic and faunal continuity does not generally exist through the Lower–Middle Cambrian boundary in Avalon (e.g., Figure 3). The upper part of the Chamberlain's Brook Formation in eastern Newfoundland and southern New Brunswick is depositional sequence 7 of the Avalonian Cambrian–Lower Ordovician (Landing, 1996a) and appears as a relatively thin, fossiliferous, limestone-bearing shale (e.g., August 12, Red Bridge Road) that often has spectacular articulated trilobites.

The middle Acadian Series is represented by the relatively thin (up to 44 m in eastern Newfoundland), richly trilobite-bearing, dark grey and black shale-dominated Manuels River Formation. Thin volcanic ashes appear through the Manuels River Formation in eastern Newfoundland and southern New Brunswick (August 12, Red Bridge Road and Manuels River; August 19, Porter Road) and in the correlative and lithologically comparable Nant-y-big Formation in North Wales (Landing, 1996a). In both North American and British Avalon, black shales dominate depositional sequence 8, and the persuasive evidence for unconformity at the base and top of the Manuels River and Nant-y-big Formations means that faunal continuity is not to be expected through the Acadian Series.

Important volcanic episodes preceded and followed development of the late Middle Cambrian depositional sequence 8–9 unconformity (discussed in next section), and a black shale-

and thin sandstone-dominated platform interval comprises latest Middle Cambrian and early Late Cambrian strata in Avalonian North America (MacLean Brook Group, Figure 3) and across Avalonian Britain (Maentwrog and Festiniog Formations in North Wales and Mancetter–Moor Wood interval in Warwickshire) (Landing 1996a). The distinction between a marginal and an inner platform disappeared at this time, and MacLean-Maentwrog-Mancetter lithofacies extend across Avalon.

Quite uniform, middle Upper Cambrian (*Parabolina spinulosa* Zone) through upper Tremadoc dark grey and black shales extend across North American and British Avalon with a loss of sand input and a possible regional deepening. This interval (Chesley Drive Group of Figure 3) is assigned to depositional sequence 10, because an unconformity is known at the contact of the *Orusia* Shales in Shropshire and most of underlying depositional sequence 9 is eroded (Landing, 1996a).

Volcanism and epeirogenic depositional sequences.—Although the upper Precambrian of the Avalonian orogen is characterized by extensive arc-related volcanism and volcaniclastic deposits, limited mixed (basaltic and rhyolitic) volcanism took place during deposition of the Avalonian cover sequence. Initial rift deposits of the upper Precambrian Rencontre Formation have significant subaerial volcanics only in the tiny area of the Cambrian–Ordovician cover sequence in northern mainland Nova Scotia (Landing and Murphy, 1991). Isolated volcaniclastic lenses at this level are known in southern Cape Breton Island (Figure 1; Landing, 1991), but are unknown in the Rencontre and Chapel Island Formations in eastern Newfoundland. Volcanism persisted into the lowest Cambrian Chapel Island Formation only in southern New Brunswick, and a U-Pb zircon age of 530 ± 1 Ma on the upper Chapel Island below the Random Formation provides an upper, but not uppermost, age on the Nemakit-Daldynian of Siberia (Isachsen et al., 1994; August 18, Somerset Street).

Upper Placentian Series rocks have not yet yielded evidence of volcanism in North American Avalon and Britain. This is unfortunate, because datable ashes would produce the first ages on Tommotian- and lowest Atdabanian-equivalent units (see Landing, 1994).

Thin K-bentonites and feldspar tuffs have been recognized in upper Branchian Series strata in southern Wales and southern New Brunswick (Landing and Westrop, 1996; August 18, Somerset Street). These ashes provide precise U-Pb zircon dates that show that the Avalonian trilobite-bearing, upper Lower Cambrian brackets an inverval from approximately 519-511 Ma and that the Lower–Middle Cambrian boundary is apparently younger than 510 Ma (Landing et al., In press).

Volcanic ashes are first documented in this volume from the uppermost Brigus Formation in St. Mary's Bay. These volcanics persist from the uppermost Branchian into the lowermost Acadian Series strata with *Acadoparadoxides* (August 15, Jigging Cove and Branch Cove); work on them with S.A. Bowring may provide the first ages that bracket the Lower–Middle Cambrian boundary. The bracketing of the Lower–Middle Cambrian boundary unconformity by volcanic ashes suggests that epeirogenic activity and associated extension may have been responsible for the development of the sequence 4–5 unconformity in eastern Newfoundland.

Basalt-dominated volcanism localized along the west St. Mary's–east Trinity axis is common through the Middle Cambrian in eastern Newfoundland. Landing (1996a) noted that pillow basalts in northern Trinity Bay (Cape Dog volcanics of Fletcher, 1972) and a volcaniclastic debris flow at Hopeall Head (Figure 2, locality Ho) are associated with the terminal stage of sequence 6 deposition. Thin K-bentonites, several of which are useful as stratigraphic markers, occur at the base of and through the Manuels River Formation in eastern Newfoundland (August 12, Red Bridge Road and Manuels River; August 17, Fosters Point) and southern New Brunswick (August 19, Porter Road), but have yielded only upper Proterozoic zircons reworked from the Avalonian orogen. These thin ashes are a precursor for locally thick pillow basalts that appear just below the top of the Manuels River Formation (Hay Cove volcanics of Fletcher, 1972) and in the overlying MacLean Brook Group

(Chapel Arm Member volcanics of McCartney, 1967). These two episodes of pillow basalt activity preceded and followed development of epeirogenic activity that led to the late Middle Cambrian unconformity between sequence 8 and 9 (August 13, Highland Cove).

Two centers of Avalonian Early(?)–Middle Cambrian volcanism occur in areas that lay on the most marginal part of Avalon. In the Bourinot Hills on the east side of the large Bras d'Or Lakes in Cape Breton Island (Figure 1), Early(?)–Middle Cambrian andesites and basaltics form a 200+ m sequence that directly overlies an uplifted block of high-grade metamorphic Proterozoic basement at the outer edge of the Avalonian marginal platform (Hutchinson, 1952; Helmstead and Tella, 1972, 1973; Landing, 1996a). The upper part of these volcanics (Gregwa Formation) is intercalated with fossiliferous, near-shore Middle Cambrian sandstones. Trilobites that occur directly below the Gregwa volcanics seem to indicate a correlation with depositional sequence 7. Our recovery of abundant zircons may allow the first U-Pb dating of depositional sequence 7 (E. Landing, S.A. Bowring, and S.R. Westrop, unpub. data).

Another volcanic center is located in the Malignant Cove autochthon, in the tiny Avalonian outlier in northern mainland Nova Scotia (Figure 1, locality NA). In this area, Landing and Murphy (1991) documented fossiliferous pebbles to giant boulders of red limestone and siliciclastic mudstone of Placentian and Branchian age in basalt granule- to cobble-clast slope deposits of the upper "Arbuckle Brook Formation." A post-Branchian, possibly Acadian age of upper "Arbuckle Brook Formation" volcanism is likely.

Upper Cambrian–lowest Ordovician volcanism is limited in Avalon, and only three thin ashes are known in this interval. A coarse-grained, feldspar-rich, upper Upper Cambrian ash at Criccieth, North Wales (Fernsides, 1910) occurs in the middle *Peltura scarabaeoides* Zone. This ash has yielded a U-Pb zircon age of 491 ± 1 Ma (Davidek et al., In press). This age is very important, because it supports Shergold's (1995) suggestion of a Cambrian–Ordovician boundary age of 490 Ma or significantly younger. Two aquagene ashes in north Wales, which lie almost on the Cambrian–Ordovician boundary and only about 2 m below the lowest occurrence of the dendroid graptolite *Rhabdinopora flabelliforme* are exposed in a cut along the Bryn-llin-fawr forestry road (Rushton, 1982), and are now under examination for a U-Pb age for the Cambrian– Ordovician boundary. Another ash, a 15 cm-thick K-bentonite, is associated with the upper Tremadoc trilobite *Peltocare rotundifrons* (Matthew) on McLeod Brook in central Cape Breton Island, Nova Scotia (Figure 1, CBI), and has yielded the first age on the Tremadoc. This 483 ± 1 Ma date (Landing et al., 1997) is somewhat older than the Tremadoc–Arenig boundary.

Field Trip Overview

From August 12–21, 1997, **Avalon 1997** participants will visit fossiliferous sections that were deposited on the marginal and inner platforms of the Avalon continent (see Landing, 1996a). Stops on August 12, 13, and 15 deal with inner platform sequences on the Avalon Peninsula, southeast Newfoundland, that lie east of or within the late Early Cambrian–Arenig St. Mary's–east Trinity axis (Landing and Benus, 1988a, 1988b; Figure 2). Cambrian sections in Conception Bay and the east side of Trinity Bay (Figure 3) and western St. Mary's Bay are incomplete in that they lack any record of the Precambrian–Cambrian boundary interval and lowest Cambrian, and the onset of deposition took place with a generalized, progressive easterly onlap of sediments referable to depositional sequences 2 (middle Placentian Random Formation), 3 (upper Placentian Bonavista Group), and 4 (Branchian Brigus Formation). The lowest local Cambrian rocks occur as fissure fills in the late Precambrian Avalonian orogen at a number of localities, and indicate that extension in a presumed transtensional regime was responsible for middle Lower–Middle Cambrian onlap and production of the St. Mary's–east Trinity axis (Landing and Benus, 1988a). Definition of the St. Mary's–east Trinity axis in the latest pre-trilobitic Early Cambrian was followed by the thickest and most stratigraphically unbroken accumulations of the siliciclastic mudstone-dominated Branchian and Acadian Series along this depocenter. The St. Mary's–east Trinity axis is associated with numerous K-bentonites in the upper Branchian Series and K-bentonites and pillow basalts in the Acadian Series.

August 16 and 17 feature sections in western Trinity Bay on the older Placentia–Bonavista axis (Figure 2). These sections (see left side of Trinity Bay column in Figure 3) lie on the Avalonian inner platform because the white quartzites and associated lithologies of the middle Placentian Random Formation constitute the oldest cover sequence rocks on the Avalonian orogen. Formation of the Placentia–Bonavista axis took place after a geologically complex history that included deposition, local block faulting and uplift, and differential erosion of the Random Formation. The thickest sections of the Bonavista Group (up to 275+ m at Keels; Figure 2, locality K) define the midline of the late Placentian Epoch Placentia–Bonavista axis. The Bonavista Group is a Tommotian–lower Atdabanian-equivalent in terms of east Siberian sections, but comprises a far thicker and a more complete succession (e.g., Landing, 1992, 1994, 1995). Unconformable onlap of the Bonavista Group took place east and west of the axis, with the lowest deposits showing a significantly younger age with greater distances to the east and west of the axis (Figure 3, see columns for S. Burin Peninsula and east Trinity Bay). Sediment-filled fissures at the base of the Bonavista Group in units as young as the Random Formation (see August 17, Sunnyside locality) also suggest onlap and deposition in a transtensional regime. Later deposition in the Placentia–Bonavista axis included relatively thin Branchian and Acadian Series units with longer hiatuses at sequence boundaries (Figure 3).

Marginal platform sections will be investigated in the Burin Peninsula, southeastern Newfoundland (August 14) and southern New Brunswick (August 18–20). Latest Precambrian–earliest Cambrian rifting or transtensional extension of the Avalonian orogen was followed by deposition of thick, relatively continuous Precambrian–Cambrian boundary interval strata in the Fortune Bay–Burin Basin of southeast Newfoundland (Figures 2, 3, see N. Fortune Bay and S. Burin Peninsula columns) and southern New Brunswick. Earliest Cambrian (i.e., early Placentian) developments in animal diversification include the appearance of diverse, behaviorally complex, metazoan trace fossil producers at the base of the Cambrian (August 14, Fortune Head global stratotype for the Precambrian–Cambrian boundary) and a later, diverse, skeletalized (aragonitic) fauna about 500 meters higher in the Chapel Island Formation (August 14, Little Dantzig Cove) (see Narbonne et al., 1987; Landing et al., 1989). A U-Pb zircon date of 530 ± 1 Ma on an upper Chapel Island Formation ash in southern New Brunswick (Isachsen et al., 1993; August 18, Somerset Street section) provides an age on the appearance of these skeletalized metazoans.

Onlap of the upper Placentian, Branchian, and Acadian Series across the marginal platform from depocenters on the inner platform (discussed above) means that the trilobite-bearing, upper Lower Cambrian through Middle Cambrian sequence is thin and unconformity-riven on the marginal plaform (see Figure 3, N. Fortune Bay and S. Burin Peninsula columns). Limestones locally comprise the basal onlap deposits of the Branchian Series (August 14, Little Dantzig Cove Brook) and Acadian Series (August 19, Hanford Brook West; August 20, Exit 113 locality), and associated faunas from these on-shore deposits are particularly rich in mollusks. Indeed, the lowest Brigus Formation at Little Dantzic Cove Brook, Burin Peninsula, southeastern Newfoundland (August 14), with a very diverse hyolith-mollusk fauna and exceptionally rare trilobites, shows that "Tommotian"-aspect faunas replaced trilobite-dominated faunas in near-shore facies. An on-shore replacement of trilobites by ostracodes and phosphatic inarticulate brachiopods is also seen in the lower facies (St. Martins Member) of the Hanford Brook Formation in southern New Brunswick (August 19, Hanford Brook West) (Landing and Westrop, 1996).

Thin volcanics in marginal platform sections in southern New Brunswick are significant in refining the geochronology of the Cambrian.

Dates of 511 + 1 Ma on the upper Hanford Brook Formation (August 18, Somerset Street, and August 20, Exit 113) support a surprisingly young Early–Middle Cambrian boundary age of less than 510 Ma and demonstrate both that the Early Cambrian is the longest part of Cambrian time and that the pre-trilobitic Early Cambrian is the longest part of the Early Cambrian (Landing et al., In press).

Biostratigraphic and Geochronologic Overview

Placentian Series

Lower Placentian Series.—The biostratigraphy of the subtrilobitic Lower Cambrian, or Placentian Series of Avalon, is based on trace and body fossils (see Landing et al., 1989; Landing, 1992). This zonation seems to reflect two stages in the Early Cambrian radiation--an earlier diversification of behaviorly complex coelomate metazoans capable of burrowing deeply through sediments, and a later diversification of skeletalized metazoans.

Low-diversity trace assemblages that include the record of sediment skimming and shallow burrowing organisms (i.e., *Gordia* Emmons, *Harlaniella* Sokolov, *Helminthoidichnites* Fitch, *Nenoxites* Fedonkin, *Palaeopasichnus* Palij, *Planolites* Nicholson) characterize the uppermost Precambrian *Harlaniella podolica* Zone of the lower Chapel Island Formation in the southern Burin Peninsula, southeastern Newfoundland. These are succeeded by the lowest occurrence of complex feeding burrows (*Trichophycus* Miller and Dyer, *Teichichnus* Seilacher, *Treptichnus* Miller) and deep dwelling burrows (*Skolithos* Haldeman, *Arenicolites* Salter) that define the base of the Cambrian (i.e., base of *Trichophycus pedum* Zone [formerly *Phycodes pedum*, see Geyer and Uchman, 1995]) at the global stratotype section and point for the Precambrian–Cambrian boundary at Fortune Head, southeast Newfoundland (August 14) (Narbonne et al., 1987).

Continuing diversification of trace producers is marked by the lowest arthropod burrows (*Rusophycus* Hall and *Cruziana* d'Orbigny) and dwelling burrows with spreiten (e.g., *Diplocraterion* Torell) about 150 m above the Precambrian–Cambrian boundary at Fortune Head (Landing et al., 1988). This assemblage defines the *Rusophycus avalonensis* Zone of the middle Chapel Island–Random Formations.

Trace producer diversification proceeded without a significant radiation in skeletalized metazoans in the Precambrian-Cambrian boundary interval of southeastern Newfoundland. Organic-walled remains of the alga *Tyrasotaenia* Gnilovskaya and the worm-like organism *Sabellidites cambriensis* Yanishevsky characterize the uppermost Precambrian-lowest Cambrian (*Harlaniella podolica*-lower *Rusophycus avalonensis* trace fossil zones) *Sabellidites cambriensis* Zone (body fossils).

The oldest skeletal (i.e., mineralized) remains are those of *"Ladatheca" cylindrica* (Grabau) in the monospecific *"L." cylindrica* Zone of the middle *R. avalonensis* Zone. Conchs of *"L." cylindrica* appear in the relatively deep shelf facies of the middle Chapel Island Formation in southeastern Newfoundland (Landing et al., 1989).

Linked lithofacies and biofacies changes related to shoaling in upper depositional sequence 1 (Figure 3) controlled the lowest appearance of diverse and abundant small shelly fossils in the upper middle Chapel Island Formation (Landing et al., 1989). *Sabellidites cambriensis* and *"Ladatheca" cylindrica* persist from lower strata, and are joined by the rostroconch *Watsonella crosbyi* Grabau, the snail *Aldanella attleborensis* (Shaler and Foerste), and the helcionellacean *Helcionella* sp. in siliciclastic mudstones at the base of the *Watsonella crosbyi* Zone in the lower part of this shoaling-upward facies. With continued shoaling and the appearance of shallow subtidal nodular and peritidal shell-hash limestones higher in the *W. crosbyi* Zone, these taxa are joined by a more diverse, facies-controlled assemblage with: 1) additional helcionelloids (*Anabarella plana* Vostokova, *Archaeospira? avalonensis* Landing, *Bemella? vonbitteri* Landing); 2) tubes of the problematica *Coleoloides typicalis* Walcott and *Anabarites korobovi* (Missarzhevsky, =*A. trisulcatus* Missarzhevsky); 3) the orthothecid *"Allatheca" degeeri* (Holm) s.l.; 4) the pseudoconodont

Maldeotaia bandalica Singh and Shukla; and 5) sclerites of the halkieriid *Halkieria stonei* Landing (Landing et al., 1989).

Rusophycus avalonensis Zone traces and *Watsonella crosbyi* Zone small shelly fossils persist into depositional sequence 2 of the upper Chapel Island Formation in southeastern Newfoundland (August 14, Little Dantzig Cove) and southern New Brunswick (August 19, Hanford Brook East). The amalgamated thick limestone traditionally assigned to the top of "member 4" of the Chapel Island Formation in the Burin Peninsula forms the base of depositional sequence 2. This peritidal limestone with a soil horizon at its base contains all of the lower *W. crosbyi* Zone small shelly fossils named above, and has the lowest occurrences of *Anabarella plana* Vostokova (helcionellacean), *Lapworthella ludvigseni* Landing and *Eccentrotheca kanesia* Landing, Nowlan, and Fletcher (tommotiids), and *Protohertzina anabarica* Missarzhevsky (pseudoconodont).

Higher strata in the upper Chapel Island Formation in southeastern Newfoundland and southern New Brunswick are higher energy, wave-dominated sandstones that lack small shelly fossils, but have diverse *Rusophycus avalonensis* Zone traces. Volcanic ashes in this interval just below the Random Formation in southern New Brunswick have yielded a 530 ± 1 Ma age on an interval correlated with the Nemakit-Daldynian Stage of Siberia (Isachsen et al., 1993).

Upper Placentian Series.—The siliciclastic mudstone-dominated Bonavista Group, which is limited to inner platform sites in southeastern Newfoundland, Cape Breton Island and northern mainland Nova Scotia (Figure 1), and eastern Massachusetts, comprises the upper Placentian Series in North American Avalon (Landing, 1996b). This interval is depositional sequence 3 of the Avalonian Cambrian (Figure 3) and includes a lower *Sunnaginia imbricata* Zone (Landing et al., 1989). The faunas of this zone are composed of most of the taxa from the older *Watsonella crosbyi* Zone but feature the lowest occurrence of the tommotiid *S. imbricata* Missarzhevsky and the phosphatic tubes of *Torellella laevigata* (Linnarsson). The hyoliths are exclusively orthothecids from this interval.

The upper Bonavista Group is referable to the *Camenella baltica* Zone. The tommotiid *Camenella baltica* (Bengtson), the oldest phosphatic inarticulate brachiopods *Obolus? groomi* (Matley) and *Micromitra phillipsi* (Holl), and hyolithid hyoliths have lowest occurrences in this zone. The conodont-like form *Hertzina elongata* Mueller, pseudoconodont *Rhombocorniculum insolutum* Missarzhevsky and Mambetov, and the bivalves *Fordilla troyensis* Barrande and *Pojetaia? terranovica* Matthew appear high in the zone.

In terms of the east Siberian successions, Tommotian Stage-equivalent strata lie within the *Camenella baltica* Zone. Faunal evidence for this correlation includes: 1) *Camenella baltica* and other bimembrate tommotiids appear in upper Tommotian-Atdabanian-equivalent intervals; 2) hyolithids and *Rhombocorniculum insolutum* appear in the middle *C. baltica* Zone and have lowest reported occurrences in the upper Tommotian; 3) *Gracilitheca bayonet*-type orthothecids appear no lower than the middle Tommotian and lower *C. baltica* Zone; 4) *Fordilla troyensis* from the uppermost *C. baltica* Zone is known elsewhere from Atdabanian-equivalent or higher strata; 5) phosphatic inarticulate brachiopods, which are unknown below the Tommotian and *C. baltica* Zone, include the *C. baltica* Zone species *Micromitra phillipsi* and *"Obolus" groomi* that may be conspecific with *Aldanotreta sunnaginensis* Pel'man from the lower Tommotian and *Cryptotreta neguerchensis* Pel'man from the middle and upper Tommotian; and 6) *Anabaritellus hexsulcatus* Missarzhevsky appears in the middle Tommotian and middle *C. baltica* Zone (Landing et al., 1989; Landing, 1991). These data suggest that much of the *Sunnaginia imbricata* Zone is older than the Tommotian, and may represent an interval that is lost at the regional Nemakit-Daldynian–Tommotian unconformity in Siberia (Landing, 1994).

Evidence on the geochronology of the upper Placentian Series is not available. Our work has not led to recognition of volcanic ashes from the upper Placentian of Avalonian North America or North Wales (i.e., lower Llanberis slates above the Red and Dorothea grits and below the green slates; Landing, 1996b).

Branchian and Acadian Series

The traditional approach to Lower and Middle Cambrian trilobite biostratigraphy of Avalonian Canada (Howell, 1928; Hutchinson, 1962) has divided the sequence into six broad zones. In ascending order, these are the *Callavia*, *Protolenus*, *"Paradoxides" bennetti*, *"P." hicksi*, *P. davidis*, and *P. forchhammeri* Zones. Fletcher (1972), in an unpublished thesis, showed that it was possible to establish a more finely divided, species-based zonation, and we agree with this conclusion. Our work on the systematics and biostratigraphy of trilobite and agnostoid faunas of the Brigus, Chamberlain's Brook, and Manuels River Formations is in progress, and this guidebook is a preliminary report.

Brigus and Hanford Brook Formations.— Restudy of the trilobite-bearing Lower Cambrian, or Branchian Series (Landing, 1992), in southeastern Newfoundland has demonstrated that *Callavia* Zone faunas (Hutchinson, 1962) are confined to the lower part of the Brigus Formation (see also Fletcher, 1972). Renamed the *Callavia broeggeri* Zone by Landing (1992), this interval is characterized by an assemblage that includes *C. broeggeri* (Walcott), *Acanthomicmacca walcotti* (Matthew), *Hebediscus attleborensis* Shaler and Foerste, *H. planus* Hutchinson, *Strenuella strenua* (Billings), *Serrodiscus bellimarginatus* Shaler and Foerste, and *Triangulaspis vigilans* (Matthew). Volcanic ashes occur in strata best correlated into the *Callavia broeggeri* Zone in the Caerfai Bay Shales, St. David's area, South Wales, and have produced a U-Pb zircon age of 519 ± 1 Ma (Landing et al., In press). This date on the stratigraphic interval with the oldest Avalonian trilobites, an interval best equated with the Botomian Stage of Siberia, supports the proposal that the majority of Early Cambrian time is pre-trilobitic (Landing, 1992, 1994; Bowring et al., 1993).

Fletcher (1972) identified *Strenuella sabulosa* Rushton from the middle of the Brigus Formation at Redlands, St. Mary's Bay (Figure 2, locality Re) and just below an interval with rare specimens of *Protolenus*. We recollected the *"S." sabulosa* interval at Redlands, and preliminary study indicates that the Redlands species may not be conspecific with *"S." sabulosa* and may be better assigned to *Myopsostrenua?* cf. *M.? sabulosa*. Our work at the Redlands section; the Hopeall Head section, east Trinity Bay; and Smith Point, west Trinity Bay (Figure 2, localities Ho and SP) shows that the lowest occurrence of this species is in a phosphate pebble lag bed or sandstone that marks the sequence 4A–4B contact within the Brigus Formation and the base of the newly named, upper member (Jigging Cove Member) of the Brigus Formation (see August 15, Jigging Cove locality; Figure 2, locality JC). This form's appearance above a sequence boundary within the Brigus marks a significant faunal change, and this middle interval of the Brigus Formation is referred to informally here as the "*Myopsostrenua?* cf. *M.? sabulosa* Faunal Interval". Our *M.?* cf. *M.? sabulosa* Faunal Interval corresponds to Fletcher's (1972) *S. sabulosa* Zone, which Fletcher reported as defining an interval in the lower Brigus Formation below strata with a low-diversity fauna with *Orodes howleyi* (Walcott); a new, undescribed species attributed to *Protolenus*; and *Catadoxides magnificus* (Walcott). This higher interval is Fletcher's (1972) *"Protolenus" howleyi* Zone.

Despite an absence of faunal data, Hutchinson (1962) assigned much of the upper half of the Brigus Formation to the *Protolenus* Zone. In its type area in the Hanford Brook Formation of southern New Brunswick, this zone is based entirely on endemic species and cannot be recognized even in other parts of Avalon. The *Protolenus elegans*-bearing interval in New Brunswick contains seven thin volcanic ashes (Landing and Westrop, 1996), and U-Pb dates on this interval include ages of 511 ± 1 Ma (Landing et al., In press). This date and the 519 ± 1 Ma age cited above on the Caerfai Bay Shale suggest that the trilobite-bearing Lower Cambrian of Avalon was only about 10 Ma long.

We have not recovered trilobites from the *Protolenus*-bearing interval in southeastern Newfoundland, but Fletcher (1972) reported a sparse, very late Early Cambrian, eodiscoid-dominated fauna with *Acidiscus theristes* Rushton and *Serrodiscus ctenoa* Rushton from higher strata of the upper Brigus Formation at the Branch Cove section, western St. Mary's Bay (Figure 2, locality

Bn). We emphasize that this interval is virtually barren, and participants on this conference will be hard-pressed to find any specimens in the greenish bands of the upper Brigus Formation. To quote Fletcher (1972, p. 187), "Collecting is by no means easy, and an individual cannot expect to amass more than two or three specimens from any one bed during the course of a full day's collecting."

An unconformity at the Brigus—Chamberlain's contact across southeast Newfoundland (Figure 3) readily defines the Lower-Middle Cambrian boundary at most localities. However, the thickest Brigus and Chamberlain's Brook sections in the St. Mary's– east Trinity axis preserve a nearly complete record of Lower–Middle Cambrian boundary interval faunas. Fletcher (1972) recovered a fauna from the uppermost 2.5 m of the Brigus Formation at Branch Cove (Figure 2, locality Bn; August 15, Branch Cove) that includes *Peronopsis* sp., a *Condylopyge*, and *Oryctocara granulata* Tchernyscheva. The first two genera and the latter species are known from the lowest Middle Cambrian Amgan Stage of Siberia (e.g., Robison et al., 1977), and represent a Middle Cambrian-aspect fauna in rocks traditionally assigned to the Lower Cambrian of the Avalon continent.

Chamberlain's Brook Formation.—Howell (1928) and Hutchinson (1962) assigned the Chamberlain's Brook Formation in Newfoundland to a single zone, the *"Paradoxides" bennetti* Zone. However, Fletcher (1972) demonstrated that *Eccaparadoxides bennetti* is restricted to the middle part of the formation and that the potential exists for defining a sequence of several species-based zones.

Of particular significance to this conference are trilobites that occur through the oldest parts of the Chamberlain's Brook Formation in the southern part of the St. Mary's–east Trinity axis at Branch Cove, west St. Mary's Bay (Figure 2, locality Bn). Fletcher (1972) reported lowest Middle Cambrian faunas with paradoxidiid fragments, a new *Condylopyge* species that persists from the underlying Brigus Formation, a new *Pagetides*, and *Braintreella* from the lower 3.3 m of the Chamberlain's Brook Formation. These taxa are joined by abundant, but uncollectable, articulated specimens of *Acadoparadoxides harlani* (Green) in slates 22 m above the base of the Chamberlain's Brook Formation (August 15, Branch Cove section).

The Lower–Middle Cambrian boundary interval in west St. Mary's Bay is significant not only for its trilobite sequence. Volcanic ashes occur through this interval at Redlands and Branch Cove, and ashes have been sampled (by EL and SRW) for U-Pb zircon dating. Ashes that have been collected include: 1) an ash at 27 m in the Branch Cove section from the upper Brigus Formation that occurs somewhat above the *Acidiscus theristes*-bearing interval; 2) an ash 5.5 m below the top of the Brigus Formation at Redlands Cove and only slightly below the *Oryctocara granulata*-bearing interval; and 3) an ash 11 m above the base of the Chamberlain's Brook Formation at Branch Cove and below the lowest occurrence of *Acadoparadoxides harlani.*

Our collecting in the Chamberlain's Brook Formation has focused on the uppermost Fossil Brook Member (Figure 3, sequence 7), which can be recognized in southeast Newfoundland and southern New Brunswick. At the type locality of the Fossil Brook Member on Porter Road, New Brunswick (August 19), our collections include *Eccaparadoxides eteminicus* (Matthew), *Acadoparadoxides acadicus* (Matthew), *Braintreella* sp., *Hartella matthewi* (Hartt), *Bailiaspis baileyi* (Hartt), and the eodiscoid *Dawsonia dawsoni* (Hartt). This fauna was assigned to the *"Paradoxides" eteminicus* Zone by Hayes and Howell (1937). *Eccaparadoxides eteminicus* is shared with southeastern Newfoundland, but other faunal similarities are masked somewhat by parochial systematics. For example, closely similar species of *Bailiaspis* and *Hartella* are present in southeastern Newfoundland (i.e., *Hartella terranovica* and *Bailiaspis manuelsensis*; see Hutchinson, 1962, and Dean and Martin, 1988), and future work may show that these are junior synonyms of species first defined in New Brunswick. Other species that occur in the *E. eteminicus* Zone in southeastern Newfoundland include *Bailiaspis venusta* Resser (= *B. prominens* Resser), *Badulesia tenera* (Hartt), and the agnostoid *Condylopyge carinata* Westergaard. In contrast to New Brunswick, eodiscoids are very rare in the Fossil Brook Member in Newfoundland, and this may reflect biofacies differences.

Manuels River Formation.—Throughout the Avalon Penninsula, the lowest *Hydrocephalus hicksi* Zone of the Manuels River Formation is defined by a sharp faunal change marked by the entry of an assemblage that includes *H. hicksi* (Salter), *Clarella venusta* (Billings), *Agraulos longicephalus* (Hicks), *"Jincella" applanata* (Salter), *Ptychagnostus fissus* (Linnarsson), *P. barrandei* (Hicks), and *Eodiscus scanicus* (Linnarsson) (see Dean and Martin, 1988). The abruptness of the faunal change across the Chamberlain's Brook– Manuels River contact is significant. Although the Chamberlain's Brook–Manuels River formational contact has long been assumed to be conformable (e.g., Hutchinson, 1962; Fletcher, 1972), differential erosion of the Chamberlain's Brook Formation took place across across southeast Newfoundland prior to Manuels River Formation deposition (Landing and Benus, 1988a; Landing, 1996b; see discussion of Manuels River and Red Bridge Road localities), and this formational contact is a sequence boundary (Figure 3).

At Highland Cove, southeastern Newfoundland (August 13), a higher fauna in the Manuels River Formation is marked by the appearance of *Meneviella venulosa* (Salter) along with *Parasolenopleura*? sp. cf. *P.*? *scania* Westergaard, a species perhaps best referred to *Jincella* Snajdr. This fauna suggests that it will be possible to divide the *Hydrocephalus hicksi* Zone into at least two subzones. These are referred to informally herein as the *Agraulos longicephalus* and *Meneviella venulosa* faunas.

The *Paradoxides davidis* Zone of the upper Manuels River Formation is defined by entry of the eponymous species along with *Anapolenus henrici* Salter, *Solenopleuropsis variolis* (Salter), and *Ptychagnostus punctuosus* (Salter). The appearance of species such as *Holocephalina americana* Resser and *Diplagnostus planicaudata* (Angelin) higher in the Highland Cove succession suggests that the *P. davidis* Zone will be divisible into subzones by future work. Post-*P. davidis* Zone faunas above the Manuels River Formation are poorly known, but some evidence exists for the latest Middle Cambrian *Lejopyge laevigata* Zone at some localities (see discussion of Manuels River section, August 12).

The Manuels River Formation at Porter Road, New Brunswick (August 19) is dominated by trilobite species different from those in southeast Newfoundland. These include *Paradoxides abenacus* Matthew and *"Jincella" acadica* (Whiteaves), and Hayes and Howell's (1937) *P. abenecus* Zone has some utility as a regional biostratigraphic unit. Presence of *Eodiscus scanicus* and the agnostoid *Ptychagnostus perrugatus* (Gronwall) indicates correlation of the *P. abenacus* Zone with the *Hydrocephalus hicksi* Zone.

Volcanic ashes at the base of and through the Manuels River Formation in southeastern Newfoundland and southern New Brunswick (August 12, 13, 17, 19) and in the lithologically comparable and correlative Nant-y-big Formation in northern Wales (see Young et al., 1994) offer the potential for U-Pb zircon ages through the upper Acadian Series. Our work with S.A. Bowring (Massachusetts Institute of Technology) has led to the recovery of well-preserved zircons. Unfortunately, all of the ashes that we have dated give ages in excess of 600 Ma, and the zircons were derived from the late Precambrian Avalonian orogen. The Manuels River Formation ashes may be basaltic tuffs.

August 12: Thin Cambrian Sections of Eastern Conception Bay

Red Bridge Road

Red Bridge Road: location.—A section through unmetamorphosed, gently east-dipping Lower–Middle Cambrian (Figure 4) is located south of the village of Kelligrews in a series of small, active quarries used for road metal and fill. The quarries are reached by automobile by turning west from Rte. 60 and traveling approximately 7 km along Red Bridge Road. The base of the measured section is north of a soccer field, and the section extends for approximately 300 m along the north side of Red Bridge Road. The Red Bridge Road

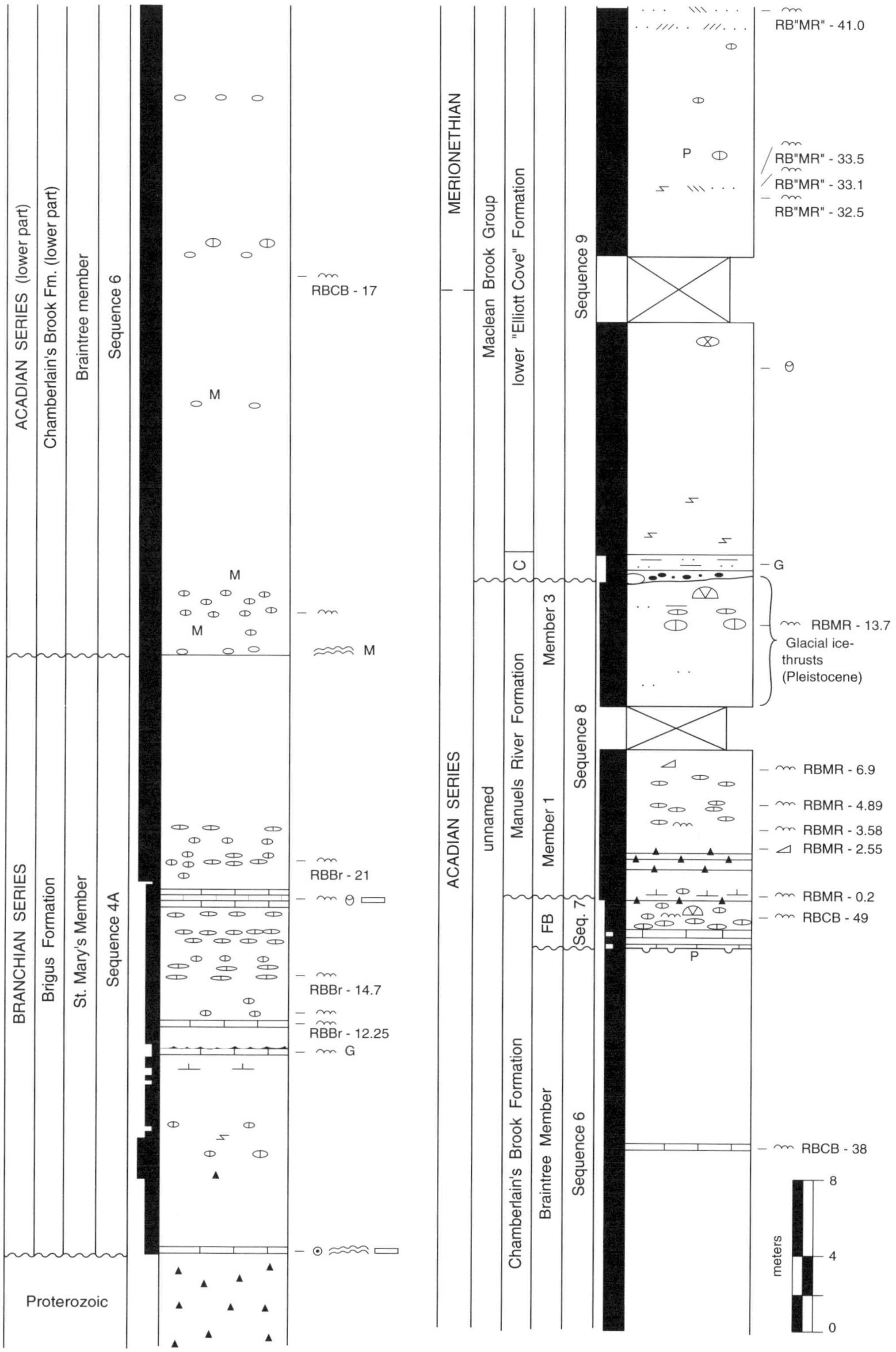

Figure 4. Section through quarries on north side of Red Bridge Road. Thickness of upper Manuels River Formation and lower MacLean Brook Group in westernmost small quarry modified by Pleistocene, ice-shove thrusts. For explanation of symbols, see Figure 6.

succession has been excavated only relatively recently, and there has been only one earlier description of the succession (Landing and Benus, 1988a).

Red Bridge Road: Brigus Formation.—The base of the Cambrian is an unconformity between hydrated, late Proterozoic volcanics and a 25 cm-thick, neomorphic, pink lime mudstone with planar cyanobacterial mats, oncoids, and wave-oriented *Coleoloides* tubes at the base of the Brigus Formation. The lower 10.75 m of the Brigus Formation consists of fossiliferous alternations of red brown, purple, and green siliciclastic mudstones with traces but no body fossils. As in other eastern Newfoundland sections, trilobites of the *Callavia broeggeri* Zone, with the eponymous species, *Strenuella strenua* (Billings), and *Serrodiscus bellimarginatus* (Shaler and Foerste), become abundant with the appearance of red shales with bedded limestones and calcareous nodules higher in the section (Figure 5). They are a major component of a massive, condensed trilobite-*Coleoloides* wacke- to packstone unit that is correlated with a comparable bed at Chapel Cove (August 13, Chapel Cove, 19.65-21.2 m above base of Brigus Formation). The Brigus Formation above the massive limestone is dominated by sparsely fossiliferous, light green to olive siliciclastic mudstones. Trilobite-bearing calcareous nodules occur in the lower 2.5 m of these green mudstones, but structureless (burrow-churned) mudstones characterize the upper Brigus Formation.

Red Bridge Road: Chamberlain's Brook Formation.—The base of the Chamberlain's Brook Formation at 31.35 m is within green siliciclastic mudstones, and is marked by manganiferous carbonate nodules and slightly higher (at 32 m), manganese-impregnated LLH mats. As noted by Howell (1925) at Manuels River, the lowest beds of the Chamberlain's Brook Formation are sparsely fossiliferous, and body fossils are limited to inarticulate brachiopods and rare, unidentifiable paradoxidid fragments. With the exception of the Fossil Brook Member, collecting in the remainder of the Chamberlain's Brook Formation at Red Bridge Road has been disappointing. The only important change from burrow-homogenized, green siliciclastic mudstone with scattered manganiferous and calcareous nodules is a 25 cm-thick trilobite packstone bed 37.65 m above the base of the Chamberlain's Brook Formation that forms the base of the third quarry west of the Precambrian–Brigus contact.

The most important lithologic and faunal changes of the Chamberlain's Brook Formation take place at the western side of the third quarry near the Chamberlain's Brook–Manuels River contact. These linked litho- and biostratigraphic changes mark the sequence 6–7 transition in Avalon (Landing, 1996a). Two rusty-weathering, 20 cm-thick trilobite packstone beds separated by 15–17 cm of green siliciclastic mudstone appear in the uppermost Chamberlain's Brook Formation. These limestones have hypichnial casts of very deep, steep-sided, horizontal burrows (up to 8 cm wide and 6 cm deep) that were produced at the sediment–water interface on overcompacted clays (compare Landing and Brett, 1987). These burrows have a fill of reworked phosphate granules and pebbles and comprise a *Glossifungites* ichnoassemblage (i.e., burrows on a submarine firm ground; e.g., Ekdale et al., 1984, and Landing, 1996a, fig. 3A, B). These limestones feature appearance of the *Eccaparadoxides eteminicus* Zone assemblage, with the lowest occurrence of the eponymous species with *Bailiaspis* and *Hartella* species (Figure 5), and are overlain by shales with abundant, large, articulated *Eccaparadoxides* specimens up to the base of the Manuels River Formation. This sequence 7 succession will also be investigated later today at Manuels River.

Red Bridge Road: Manuels River Formation.—The lower 8.4 m of the Manuels River Formation is exposed in the high wall at the west end of the third quarry. The base of the Manuels River Formation is a 2 cm-thick K-bentonite at the foot of the high wall and at the base of the somewhat darker siliciclastic mudstones (see discussion of base of Manuels River Formation in Landing and Westrop, this volume, and at the Manuels River stop). Very thin K-bentonites occur through the lower Manuels River Formation (i.e., 1.6, 2.25, and 2.45 m above the basal K-bentonite). Trilobites (*Hydrocephalus hicksi* Zone) of the lower 8.4 m in this quarry occur in thin storm-lag lenses in silici-

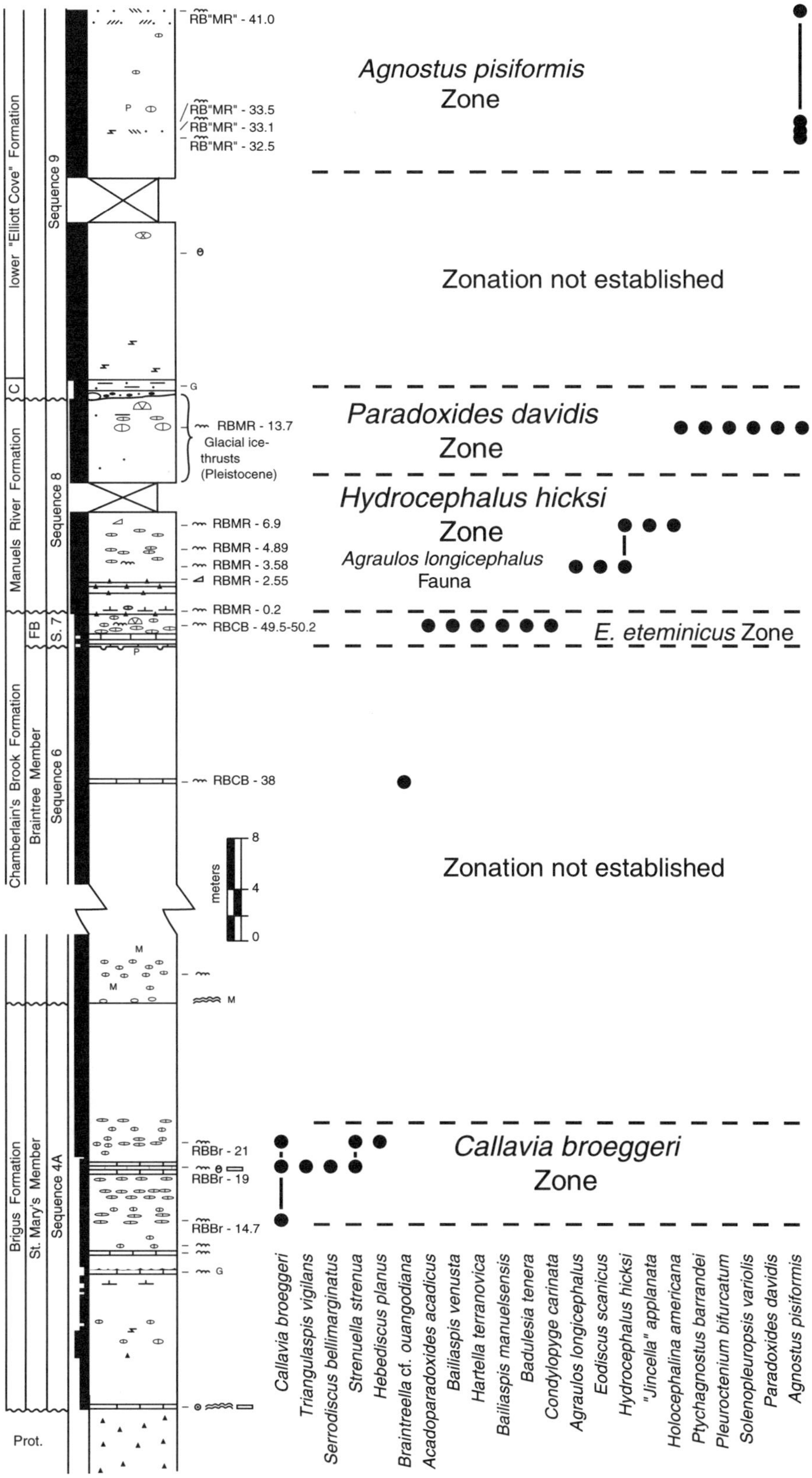

Figure 5. Ranges of trilobites and agnostoids at Red Bridge Road. See Figure 4 for complete section log.

clastic mudstone, and were collected 3 m, 3.58 m, and 6.9 m above the base of the Manuels River Formation. Excellent specimens of a hyolithid species with conchs associated with opercula and helens are present 2.55 m above the base of the Manuels River Formation (in preparation by EL).

The lower Manuels River Formation at the third quarry is capped by glacial drift, and the section continues above a covered interval just to the west in a fourth small quarry. This last section is important, as it shows the upper contact of the Manuels River Formation and previously unreported stratigraphic developments at the Manuels River Formation–MacLean Brook Group contact. The lower part of the fourth quarry is composed of dark grey to black shale of the Manuels River with numerous, Pleistocene, ice-shove thrusts with crush zones. *Paradoxides davidis* Zone trilobites, with *P. davidis* Salter, *Ptychagnostus punctuosus* (Salter), *Holocephalina americana* Resser, and *Solenopleuropsis variolis* (Salter), occur in large calcareous nodules 13.7 m above the base of the Manuels River (Figure 5). Articulated *P. davidis* specimens appear in shale 60 cm below the top of the formation. This succession of large calcareous nodules and overlying articulated *P. davidis* will be seen at Manuels River, and characterizes member 3 of the Manuels River Formation (Landing and Westrop, this volume; August 13, Highland Cove).

Red Bridge Road: MacLean Brook Group and sequence 8–9 unconformity.—The Manuels River is sharply overlain by a brownish-weathering, 30–50 cm-thick conglomeratic sandstone with phosphatic pebbles to 3x4x2 cm in diameter. This conglomeratic sandstone, a somewhat thicker version of that seen at the top of the Manuels River at the Manuels River section later today, is overlain by a 60 cm-thick, thin-bedded, weakly calcareous, green-grey glauconitic sandstone with *Teichichnus* burrows and other horizontal traces and then by dark-grey micaceous shale with the lower Upper Cambrian agnostoid *Agnostus pisiformis* at 16.7–17.7 m and 25.2 m above the Manuels River Formation.

These developments of a conglomeratic sandstone and overlying glauconitic sandstone appear above the Middle Cambrian *Paradoxides davidis* Zone, and mark the late Middle Cambrian sequence 8–9 unconformity known across North American and British Avalon (Landing, 1996a). The uppermost Middle Cambrian biostratigraphy from the top of the Manuels River Formation into the lowest MacLean Brook Group is discussed below for the Manuels River section. Although the glauconitic sandstone does not appear at Manuels River, it is correlative with the the massive sandstones that overlie the Manuels River further west at Highland Cove in eastern Trinity Bay (August 13). This new unit at the top of the Manuels River Formation, inclusive of the conglomeratic sandstone and overlying glauconitic sandstone, is termed the Cavendish Formation (new, Landing and Westrop, this volume), a unit that reaches up to 45 m in thickness in southern St. Mary's Bay and correlates with sandstone facies that overlie *Paradoxides davidis* Zone black shales in Warwickshire, England (i.e., Mancetter grits and shales) and north Wales (Maentwrog Formation) (Landing, 1996a).

Manuels River bridge overlook

Manuels River bridge: location.—After returning to Rte. 60 from the Red Bridge Road quarries, field trip participants will drive northeast and stop near the bridge over Manuels River in the village of Manuels River. A brief inspection will be made of the gently north-dipping Lower–lowest Middle Cambrian section just below the bridge.

Descriptions of the sequence below the highway bridge across Manuels River in the village of Manuels River are available in Dale (1915), Howell (1925), Anderson (1987), and Landing and Benus (1988a). The bedrock under and upstream from the bridge is late Proterozoic Holyrood granite with inclusions of Harbour Main Group volcanics (Anderson and King, 1982; see discussion of Duffs section, below, for geologic and geochronologic significance of the Holyrood granite).

Manuels River bridge: lower conglomerate.—The Holyrood granite is overlain by 5.5 m of conglomerate composed largely of rhyolite cobbles and pebbles in a sandstone matrix. Large rhyolite boulders up to 3.6 m in diameter (Dale, 1915, p. 377) overlie the Holyrood granite. Although Anderson (1987) regarded this conglomerate as a local "beach deposit" referable to the Brigus Formation, absence of fossils and restricted distribution of this conglomerate (i.e., with outcrop limited to Manuels River and isolated outcrops 150 m to the north [Howell, 1925, fig. 3]) preclude a stratigraphic assignment or an understanding of the depositional significance of the conglomerate. Alternative explanations include deposition with:

1) faulting and extension associated with Random Formation onlap (middle Placentian);
2) development of the late Placentian Placentia–Bonavista axis and accumulation of the lower–middle Bonavista Group; or
3) growth fault activity with development of the St. Mary's–east Trinity axis, and accumulation of the late Placentian Fosters Point Formation elsewhere in the Avalon Peninsula (see Figure 2).

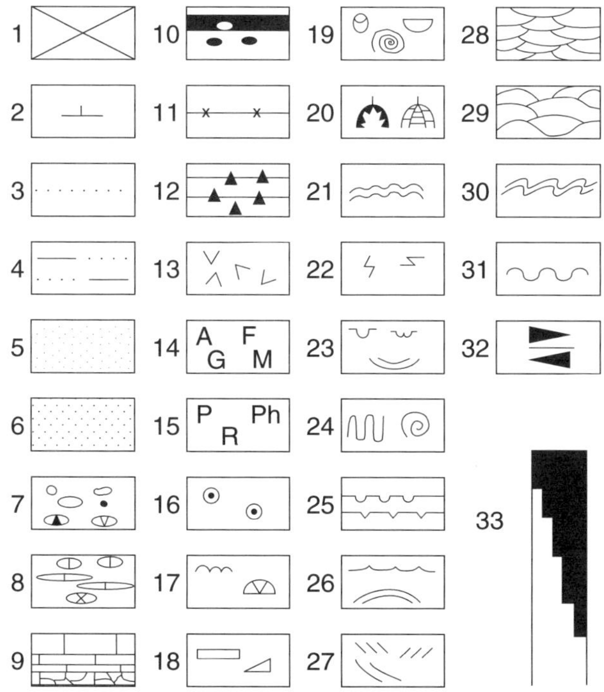

←
Figure 6. Symbols in stratigraphic columns. 1, covered interval; 2, siliciclastic mudstone or shale, often slatey, may have calcareous laminae (symbol in center of rectangle); 3, thin sandstone bed in shale or siliciclastic mudstone; 4, shale and thin sandstones; 5, thin- to medium-bedded sandstones, typically fine- to medium-grained; 6, coarse-grained sandstones; 7, conglomerate, may have phosphatic (black), volcanic (with triangle), or granitic (letter "v") clasts; 8, calcareous nodules (top ellipses) that may form coalesced nodule beds (middle symbols) and sideritic nodules (ellipse with "x"); 9, limestone bed (upper) and nodular limestone bed (lower); 10, phosphatic bed, may have reworked clasts (white ellipse) and phosphatic nodules (black ellipses); 11, siliceous (cherty) siliciclastic mudstone with volcaniclastic component; 12, massive volcanic ash bed (upper triangles) and thin ash beds (triangles on line); 13, granitic rock of Avalonian orogen; 14, arkose (A), feldspathic (F), glauconitic (G, typically metamorphosed to chlorite), hematitic (H), manganiferous (M); 15, pyrite (P), phosphatic (Ph), rhodochrosite (R); 16, oncoids or pisolites; 17, disarticulated trilobites (upper left) and complete carapaces (lower right); 18, conoidal problematica (upper) and hyoliths (lower); 19, brachiopods (upper left), ostracodes (upper right), mollusks (lower); 20, anisograptid dendroids (left) and *Rhabdinopora flabelliformis*; 21, stromatolites; 22, *Planolites* (left) and horizontal traces; 23, *Arenicolites* (upper left), *Didymaulichnus* (upper right), and *Teichichnus* (lower); 24, *Taphrhelminthoida* (left) and new genus reported as *Cylindrichnus*; 25, burrow hyporeliefs (upper) and dessication cracks (lower); 26; wave ripples (upper) and hummocky cross-stratification (lower); 27, microcross lamination (upper) and trough cross-beds (lower); 28, trough cross-bedding; 29, gravel dunes; 30, contorted (slumped) bedding; 31, unconformity; 32, fault, shows orientation to bedding and offset direction; 33, rock color code shows six steps from black (top) and successively down through medium or dark grey; green; red or pinkish; purple red or purple; and white, buff, or yellow (bottom).

Manuels River bridge: Brigus–lowest Chamberlain's Brook Formations.—An equivalency of the conglomerate with units older than the Brigus Formation is suggested by slabbed sections of a 25 cm-thick, grey-green, calcareous sandstone that overlies the conglomerate and is definitely referable to the lower Brigus Formation. This sparsely fossiliferous *Lingulella*-bearing sandstone, which was reported by Howell (1925, p. 25) as a "shaly, pyritiferous, pink and blue limestone containing great numbers of what are probably pteropod shells", has angular phosphatic clasts to several centimeters in diameter; limonite-impregnated oncoids and circumrotational stromatolites; and bored limestone clasts with phosphatic rinds. In these features, it resembles lag sandstones that overlie unconformities in other parts of the Avalonian Lower Cambrian (e.g., Landing and Benus, 1988b).

Green siliciclastic mudstones of the Brigus Formation are poorly exposed 1–2.5 m above the conglomerate. Above a covered interval, an important stratigraphic development is marked by red siliciclastic mudstones 14–15 m above the conglomerate that are overlain by a 10 cm nodular lime mudstone and an overlying greenish, 30 cm-thick intraclast wackestone with phosphatized clasts, lingulelloid brachiopods, and trilobite fragments. This nodular lime mudstone–intraclast wackestone unit forms a broad pavement at the first easterly bend of Manuels River below the bridge. We have not been able to collect identifiable trilobites from this limestone. Dale (1915, p. 380) reported this development of red siliciclastic mudstone and limestone somewhat below the characteristic Chamberlain's Brook Formation facies of green siliciclastic mudstone with manganiferous nodules.

The Brigus–Chamberlain's Brook contact and unconformity at the Lower–Middle Cambrian boundary lies only 1.3 m above the top of the phosphatic limestone. This formational contact and boundary between the Lower Cambrian Subsystem and Acadian Series is marked by the base of a 10 cm-thick, manganiferous, stromatolitic green mudstone bed exposed at the easterly bend in Manuels River. Howell (1925, beds 1–35) described a complete section through the Chamberlain's Brook Formation further downstream along the banks of Manuels River. Our experience is that the bed-by-bed description of Howell's (1925) section is not readily applicable to the Chamberlain's Brook Formation in Manuels River. His reported thickness of the formation (68.4 m) is too great, and the thickness of the Chamberlain's Brook Formation here is closer to the 51 m at the nearby Red Bridge Road section (Figure 4).

Manuels River bridge: sequence 4A–4B contact in Brigus Formation.—The phosphatic intraclast limestone bed noted above is interpreted to overlie a newly recognized stratigraphic break in the Brigus Formation that corresponds to the sequence 4A–4B boundary and to the boundary between the St. Mary's and Jigging Cove Members of the Brigus Formation (see Westrop and Landing, this volume; compare with Brigus sections at Jigging Cove [August 15] and Smith Point [August 16]).

An abrupt replacement of *Callavia broeggeri* Zone trilobites by those of the *Myopsostrenua?* cf. *M.? sabulosa* Faunal Interval takes place at the sequence 4A–4B unconformity at a number of localities (e.g., a trilobite packstone with phosphatic clasts that is Hutchinson's [1962, p. 133] locality 20402 at Hopeall Head, eastern Trinity Bay, and immediately above a phosphate pebble bed 58 m above the base of the Brigus Formation at Redlands, St. Mary's Peninsula [see Figure 2, localities Ho and Re]). These stratigraphic relationships pose problems for the utility of part of the Avalonian Lower Cambrian trilobite zonation as it has been used by a number of workers.

"Catadoxides magnificus *Zone*".—Trilobites are not well known from the Brigus Formation at Manuels River. The only form that has been reported is the poorly known and inadequately illustrated *Catadoxides magnificus* (Matthew). This species was suposedly collected from green shales low in the Brigus section (Dale, 1915; Howell, 1925). This places the provenance of the type material of the species in sequence 4A and in the *Callavia broeggeri* Zone. Fletcher (1972) subsequently reported, but did not illustrate, *C. magnificus* from one horizon in much higher strata of depositional sequence 4B from the *Protolenus*

Zone at Redlands, at the headland between Placentia and St. Mary's Bays (Figure 2).

Fletcher (1972) proposed a "*Catadoxides magnificus* Zone" in an unpublished dissertation, and subsequently used it (Bengtson and Fletcher, 1983) as a zone-level unit for middle Branchian Series strata of the Brigus Formation well above the *Callavia broeggeri* Zone. Although used by subsequent workers as a legitimate Avalonian trilobite zone (e.g., Geyer and Palmer, 1995), the "*Catadoxides magnificus* Zone" is actually a nonfossiliferous, barren interval bounded below by a *Protolenus* Zone and above by the *Acidiscus theristes* Zone on the basis on Fletcher's (1972) data (Landing, 1992). Occurrences of *C. magnificus*, which were not documented stratigraphically by Fletcher (1972), supposedly occur in the trilobite zones below and above the barren interval, and a biostratigraphic "zone" without any fossils was named for the interval by Fletcher (1972). Available biostratigraphic control on the sequence 4A–4B boundary emphasizes the futility of continued efforts to utilize a "*Catadoxides magnificus* Zone" in Avalon. Indeed, the species seems to occur as low as the lower Branchian *Callavia broeggeri* Zone of the St. Mary's Member of the lower Brigus Formation at Manuels River and ranges high into the Jigging Cove Member (see Landing and Westrop, this volume).

A final caveat to the use of a "*Catadoxides magnificus* Zone" is that the type specimen of this species may prove to be a poorly preserved *Callavia broeggeri* specimen. This interpretation is consistent with its first description from the St. Mary's Member at Manuels River.

Manuels River section

Manuels River section: location.—At the end of the Manuels River Bridge stop, drive north on Rte. 60 past the school on the left and continue for approximately 2 km. Turn left (west) on Iveney Place and park at the end of the street. Directly opposite (west) of the parking area and across Manuels River is a gently northwest-dipping, long section that extends from the upper Chamberlain's Brook Formation (lower Middle Cambrian) through the lower Upper Cambrian (*Parabolina spinulosa* Zone) of the upper MacLean Brook Group (Hutchinson, 1962; Landing, 1996a). Descend the stairs to Manuels River, and walk approximately 150 m upstream. In the course of this walk, participants will pass a low cliff of black shales on the east side of Manuels River that includes the Manuels River Formation– MacLean Brook Group unconformity (discussed below). Bergstrom and Levi-Setti (1978) examined the phenotypic variation of *Paradoxides davidis* through the upper 4 m of the Manuels River Formation from an excavation at the west end of this cliff. The field trip participants will ford Manuels River to a relatively narrow (5–10 m wide), sloping cliff section framed by spruce trees that extends from the upper Chamberlain's Brook Formation into the lower MacLean Brook Group (Figure 7; see photographic illustration by Howell, 1925, fig. 2). This locality includes the type section of the Manuels River Formation (Hutchinson, 1962). The Manuels River section has been given a protected status by the Government of Newfoundland and Labrador, and visitors must not collect.

Manuels River section: type section of Manuels River Formation.—The uppermost Chamberlain's Brook Formation–lower MacLean Brook Group section complements the Red Bridge Road section. Fletcher's (1972) unpublished restriction of the Manuels River Formation is followed herein (see Landing and Westrop, this volume). Hutchinson (1962) included Howell's (1925) beds 26–125 in the type section of the black and dark grey siliciclastic mudstone-dominated Manuels River Formation. This definition included an interval of greenish siliciclastic mudstones characteristic of the Chamberlain's Brook Formation with "*Paradoxides*" *bennetti* Zone fossils in the lower Manuels River Formation. A more appropriate and regionally correlatable base of the Manuels River Formation is Fletcher's (1972) use of Howell's (1925) bed 36, a whitish K-bentonite, as the base of the Manuels River Formation. Presumably, Hutchinson's (1962, p. 22) definition of bed "26" as the base of the Manuels River Formation was a typographic error for bed "36."

Manuels River section: Chamberlain's Brook–Manuels River unconformity.—The Chamberlain's

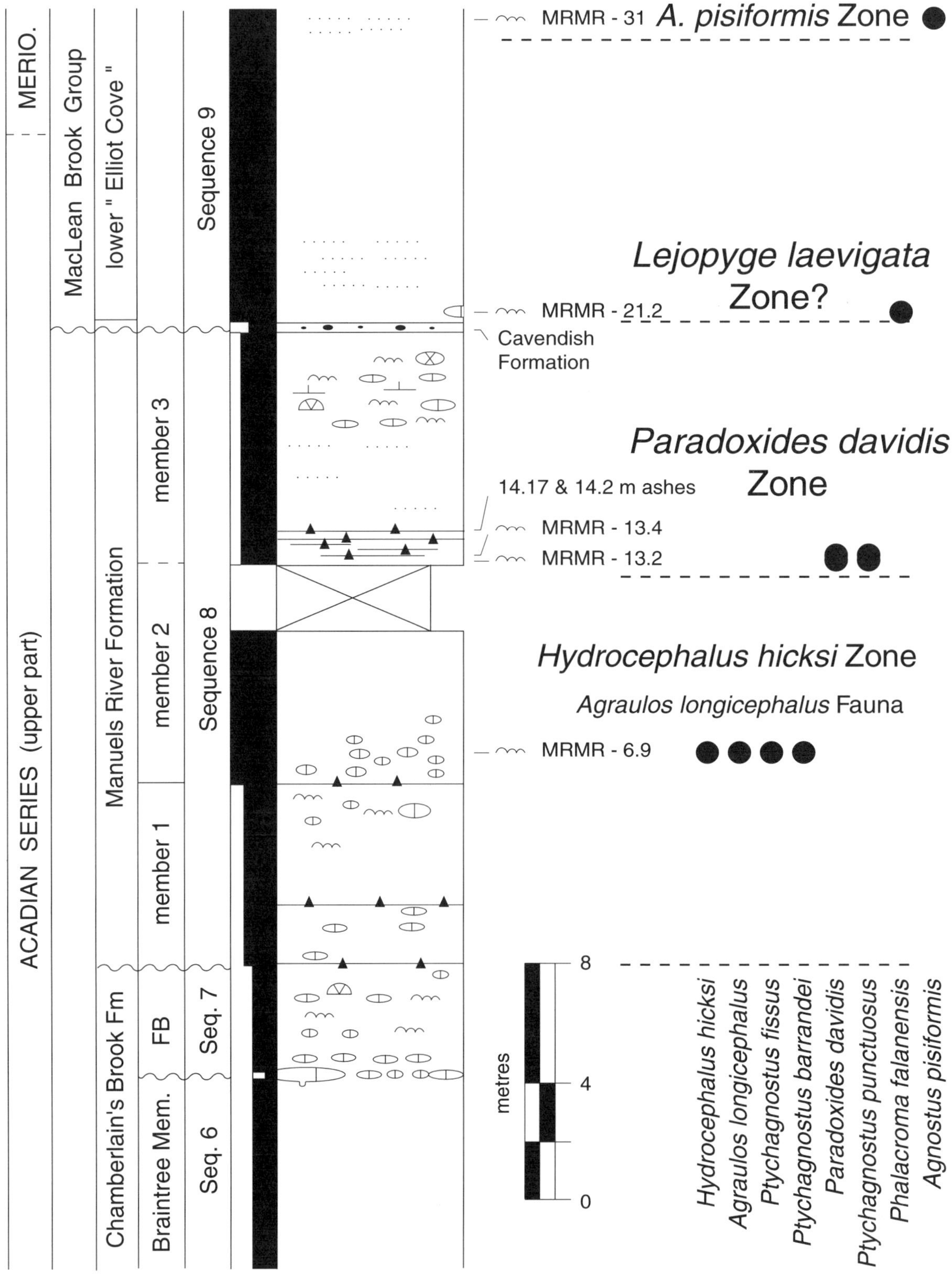

Figure 7. Middle Cambrian section and ranges of trilobites and agnostoids at type section of Manuels River Formation on Manuels River. For explanation of symbols, see Figure 6.

Brook-Manuels River contact has traditionally been regarded as conformable (Howell, 1925; Hutchinson, 1962; Fletcher, 1972; Williams et al., 1985, p. 231). However, use of the lowest K-bentonite to define the base of the Manuels River Formation (Fletcher, 1972) helps demonstrate the Chamberlain's Brook–Manuels River unconformity originally proposed by Landing and Benus (1988a) and subsequently regarded as the depositional sequence 7–8 unconformity in Avalon (Landing, 1996a). The base of the lowest of the two nodular limestone beds that occur in the lowermost Fossil Brook Member of the Chamberlain's Brook Formation (i.e., beds 21 and 24 of Howell, 1925) is 3.4 m below the K-bentonite at Manuels River. This stratigraphic distance is only 2.2 m at Red Bridge Road, and several nodular limestones at the top of the Fossil Brook Member at Manuels River are absent at Red Bridge Road. The amount of stratigraphic cut-out of the upper Chamberlain's Brook Formation prior to Manuels River deposition is even greater on the east side of Trinity Bay at Hopeall Head (Figure 2), where the Fossil Brook Member is restricted to a limestone bed (Hutchinson, 1962, p. 132, bed 32) that is overlain directly by the Manuels River Formation. These stratigraphic details indicate that a complete biostratigraphic transition from the Chamberlain's Brook Formation into the agnostoid-bearing *Hydrocephalus hicksi* Zone of the lowest Manuels River is not present in eastern Newfoundland. In addition, the presence of a K-bentonite at the base of the Manuels River Formation at many localities (see stops on August 13, Highland Cove, and August 17, Fosters Point) is explainable as a result of a break in background deposition of siliciclastic mudstone during onlap of the Manuels River "sea" and accumulation of a bentonite on a surface of non-deposition.

Manuels River section: Manuels River Formation stratigraphy and faunas.—Three of the four members of the Manuels River Formation are present at Manuels River (see Landing and Westrop, this volume, and stops on August 13, Highland Cove, and August 17, Fosters Point). The somewhat lighter colored (dark grey) siliciclastic mudstone of the lowest Manuels River Formation at Manuels River is regionally recognizable and is bounded by a thick (7 cm) ash at the base of the Manuels River Formation and a thinner ash (2–4 cm) at 5.9 m. This interval is member 1 of the Manuels River Formation and includes a thin ash (2–3 mm) at 1.9 m.

Member 1 is overlain by laminated, very dark grey siliciclastic mudstones with trilobites locally abundant in wackestone nodules of member 2. The lower part of this dark grey interval is relatively unfossiliferous, but yielded a *Hydrocephalus hicksi* Zone fauna (*Agraulos longicephalus* Fauna) at 6.9 m (Figure 7). These very dark grey mudstones are followed by a covered interval (11–13.2 m) and then by an interval with mm-thick ashes (at 13.5, 13.65, 13.8, and 14.05 m) that are assigned to member 3. A much thicker amalgamated ash composed of four ashes with black shale laminae appears slightly higher in member 3 (14.17–14.2 m), and is followed by an interval with articulated *Paradoxides* specimens at 18 m. This amalgamated ash is traceable into western Trinity Bay to Fosters Point (August 17, Fosters Point), where it is also overlain by an interval with articulated *Paradoxides* specimens. These upper ashes and higher strata with articulated *Paradoxides* specimens form member 3 of the Manuels River Formation.

Howell (1925) reported a transition from the *Hydrocephalus hicksi* Zone into the *Paradoxides davidis* Zone at 11 m. Our collections from the interval with mm-thick ashes (13.5–14.05 m) have a *P. davidus* Zone fauna with *P. davidis* and *Ptychagnostus punctuosus*. Articulated *Paradoxides davidis* specimens can be observed in float along a bench at about 18 m in the measured section. This bench was excavated by commercial collectors and lies in the interval that Bergstrom and Levi-Setti (1978) did their biometric study of *P. davidis* on the east side of Manuels River (noted above).

Manuels River section: sequence 8–9 unconformity.—The top of the Manuels River Formation and *Paradoxides davidis* Zone is marked by a thin (8–18 cm), yellowish-weathering, fine grained sandstone with angular to subrounded siliciclastic mudstone and phosphatic clasts up to several centimeters in diameter (Howell, 1925, bed 125). Hutchinson (1962) earlier included bed 125 as the top layer of the Manuels River Formation type section. This bed has been regarded as providing evidence for a hiatus, and Walcott (1900b) and Hutchinson (1962) considered that it marked the Middle–Upper Cambrian boundary, even though the lowest specimens of the lowest Upper Cambrian agnostoid *Agnostus pisiformis* (Linnaeus) occur 10 m higher (Howell, 1925).

Poulsen and Anderson (1975) recovered a sparse, uppermost Middle Cambrian fauna from the lower 1.5 m of the siliciclastic mudstones above the conglomeratic sandstone. They assigned this fauna to the *Lejopyge laevigata* Zone. Specimens that they identified as *Peronopsis insignis* Wallerius (Poulsen and Anderson, 1975, pl. 2, figs. 10–13) are in our opinion too poorly preserved for a confident identification. We agree with Rushton (1978) and Robison (1994) that Poulsen and Anderson (1975, pl. 2, figs. 1–9) incorrectly identified *Lejopyge laevigata* (Dalman) in their collections and follow Rushton in transferring them to *Phalacroma falanensis* Westergaard (= *P. bairdi* Hutchinson; see Rushton, 1978, p. 257), a species which we have recoved 0.5 m above the conglomeratic sandstone (Figure 7). According to Rushton, *P. falanensis* ranges from the *Lejopyge laevigata* Zone into the Upper Cambrian. Poulsen and Anderson (1975) also recorded *Andrarina costata* (Angelin) from this interval, a species confined to the *Lejopyge laevigata* Zone in Sweden (Westergaard, 1948). Thus, despite misidentifications, the evidence supports Poulsen and Anderson's (1975) biostratigraphy, and the sequence 8–9 unconformity lies within the Middle Cambrian.

Work on the stratigraphy of the Manuels River Formation indicates a hitherto unappreciated depth of erosion on the Manuels River prior to deposition of the MacLean Brook Group. As noted at Highland Cove (August 13), a higher, approximately 6 m-thick interval with articulated *Paradoxides* specimens (i.e., member 4 of the Manuels River Formation) is recognized on the east side and south end of Trinity Bay (e.g., Highland Cove and MacLeod Point east; see Figure 2, localities HC and MPE). The absence of member 4 of the Manuels River Formation at Manuels River and Red Bridge Road is attributed to erosion of the upper Manuels River Formation. As suggested by Dean and Martin (1988, fig. 4), the lowermost two zones of the *Paradoxides forchhammeri* Stage (*Goniagnostus nathorsti* and *Solenopleura brachymetopa* zones) may be missing at Manuels River.

This unconformity at the base of Howell's (1925) bed 125 marks the sequence 8–9 boundary in Avalon (Landing, 1996a), and the conglomeratic sandstone is the lowest deposit of sequence 9 and the MacLean Brook Group. The sandstone is no longer regarded as the top of the Manuels River Formation (fide Hutchinson, 1962), and is the feather edge of the Cavendish Formation (Landing and Westrop, this volume; see August 13, Highland Cove).

August 13: Uppermost Placentian–Acadian, Eastern Conception and Western Trinity Bays

Duffs

Duffs: location.—The upper Proterozoic–Lower Cambrian at Duffs (Figure 8) lies on the southeastern shore of Conception Bay several hundred meters north of the oil tanker dock. The section is reached from Rte. 60 by driving west for several hundred meters down the access road that leads to the power plant at Duffs and turning left onto Duffs Road. Vehicles can be parked at the unpaved end of Duffs Road, where large boulders have been placed to block access to the power plant road (asphalt) that parallels the railroad tracks. A short path and the power plant road lead west to an elevated pipeline and the railroad tracks. The route continues north along the railroad tracks for about 150 m. Participants will descend a steep, treeless slope to the shore of Conception Bay. The unmetamorphosed, gently north-dipping Lower Cambrian section at Duffs forms a reentrant north of the resistant low cliffs of the Holyrood granite.

Duffs: Holyrood granite.—The Proterozoic Holyrood granite at the base of the section is important in understanding the origin of the Avalon terrane because geological relationships and geochronology show it to be one of the oldest units in Avalonian Newfoundland. The Holyrood has been considered co-magmatic with the oldest volcanic rocks of the Avalon Peninsula (i.e., the Harbour Main Group; see Hughes and Bruckner, 1972). Because clasts of Harbour Main volcanics occur in lower Conception Group siliciclastics (McCartney et al., 1966) well below the level of the ca. 565 ± 3 Ma year old, soft-bodied Ediacaran fossils of the Mistaken Point Formation (Benus, 1988), the Harbour Main-Holyrood complex records volcanism and intrusion much before the late Proterozoic Avalonian orogeny. A wide range of late Proterozoic ages have been reported from the Holyrood granite. An early Rb-Sr date of 585 ± 15 Ma (McCartney et al., 1966) has been revised by Dalmeyer et al. (1981) and Conway Morris (1988) to 590 ± 11 Ma and 594 ± 11 Ma, respectively, on the basis of the new decay constant. Perhaps more significant are the much older U-Pb zircon ages of 620 +2.1/-1.8 Ma (Krogh et al., 1983) and 610 Ma (King et al., 1985) on the Holyrood.

Duffs: Fosters Point Formation.—General stratigraphy of the Duffs section and features of the nonconformity between the late Proterozoic "basement" of the Holyrood granite and Lower Cambrian have been discussed by Anderson (1987) and Landing and Benus (1988a). Sediment-filled cracks up to 0.9 m-deep in the Holyrood granite have been cited as evidence for regional extension in a transtensional regime and foundering that led to Placentian transgression and definition of the St. Mary's–east Trinity axis (Figure 2) on the inner platform during the latest Placentian–early Branchian (Landing and Benus, 1988a, 1988b; Landing, 1996a).

Anderson (1987) assigned the entire section above the Holyrood granite to the Brigus Formation. However, the basal stromatolitic and locally conglomeratic phosphatic bed (0.0–0.8 m-thick, depending on local relief on the Holyrood granite) and the overlying 0.5 m-thick, condensed, greyish wacke- to grainstone limestone are referable to the Fosters Point Formation (Landing and Benus, 1988b). As will be seen at the next stop at Chapel Cove, the typically reddish limestones of the Fosters Point Formation are replaced by greyish, somewhat phosphatic and

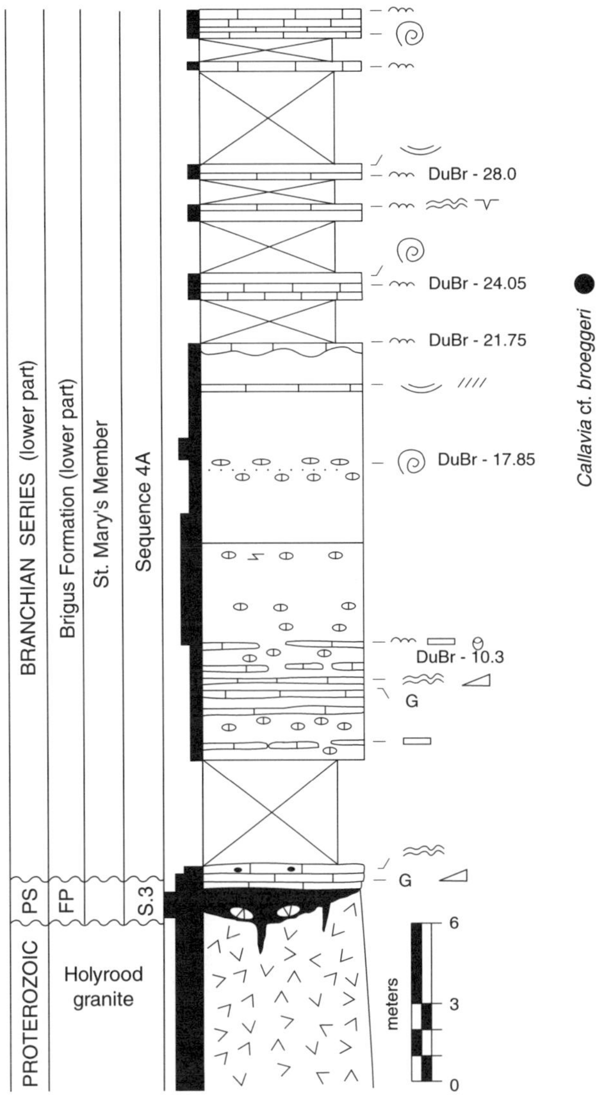

Figure 8. Upper Proterozoic and uppermost Placentian–lower Branchian section at Duffs. For explanation of symbols, see Figure 6.

dolomitic limestone at the southeastern margin of Conception Bay. This greyish limestone at Duffs yields a characteristic uppermost Placentian Series (upper *Camenella baltica* Zone) assemblage with the tommotiid *C. baltica* (Bengtson), rostroconch *Watsonella crosbyi* Grabau, snail *Aldanella attleborensis* (Shaler and Foerste), large "helcionellid" *Randomia auroriae* Matthew, and strongly transversely ribbed *Hyolithellus* conchs. The significance of this uppermost Placentian fauna is that it demonstrates that almost all of the sub-trilobitic Lower Cambrian is absent on the easternmost preserved part of the Avalonian inner platform in southeastern Newfoundland.

Duffs: Brigus Formation.—The top of the Fosters Point Formation has a thin phosphate- and limonite-impregnated cyanobacterial mat as a cap. The basal bed of the Brigus Formation is a 0.4 m-thick, green-grey, *Lingulella*-rich neomorphic limestone with oncoids and fragments of the cyanobacterial mat (Landing, 1996a, fig. 4b). This limestone is the first deposit on the Placentian–Branchian series sequence boundary. Unlike the basal pinkish, trilobite hash-rich beds of the Brigus Formation that rest unconformably on the Fosters Point Formation on the west side of Conception Bay and in Trinity Bay (see Smith Point, August 16), the lowest green limestones and siliciclastic mudstones of the Brigus in southeastern Conception Bay are a Branchian facies that lacks trilobites. The overlying green and red siliciclastic mudstone facies through the lower 20 m of the section at Duffs is sparsely fossiliferous, with shelly faunas dominated by *Coleoloides* tubes and hyoliths in thin nodular limestones capped by thin cyanobacterial mats. These limestones regularly contain abundant *Lingulella* valves, a form unknown in the Placentian Series, which appears with *Callavia broeggeri* Zone trilobites elsewhere in eastern Newfoundland.

Trilobites have their lowest occurrence and suddenly dominate the fossil pack- to grainstone caps of the thicker limestones above 20 m at Duffs. However, most of the material consists of fragments that cannot be identified even to the generic level. Our collections yielded *Callavia* sp. cf. *C. broeggeri* Walcott at 24.05 m (Figure 8). The appearance of episodically higher energy facies, a red-dominated rock color, disappearance of cyanobacterial caps on limestones, and low frequency of *Lingulella* valves in this part of the section suggest that depositional environments changed enough to allow the local appearance of a trilobite-dominated biofacies.

At the end of the stop, the Chapel Cove locality will be pointed out to the southwest across Conception Bay. The proximity of this section to Duffs and the appearance of older facies of the Cuslett Formation below the Fosters Point Formation emphasize the significance of the St. Mary's–east Trinity axis as a depocenter.

Chapel Cove

Chapel Cove: location.—The shore section at Chapel Cove Point (Figure 9) lies 4 km southwest of Duffs across the south end of Conception Bay. It is reached by driving to the village of Chapel Cove ("Chapels Cove" on newer road maps) and continuing approximately 1 km NNE along the coast road that follows the west side of Chapel Cove Point. Vehicles should be parked where the road swings northwest and begins to climb the ridge. Walk approximately 0.5 km in pastures along the coast to Chapel Cove Point. The point features a gently south-dipping nonconformity between the Holyrood granite and Lower Cambrian.

A section through the upper Placentian Series (uppermost Cuslett Formation and Fosters Point Formation) and into the lowermost Brigus Formation can be measured in gently south-dipping strata (Figure 9). The section becomes steeply dipping immediately west of the point. A section through to the base of a thick, vertical red trilobite limestone (20.7–21.2 m) that forms the west side of Chapel Cove Point is completely exposed along the shore. This thick red limestone forms a broad, gently east-dipping platform 150 m to the northwest of Chapel Cove Point. The upper 13.35 m of the section (Figure 9, $CCBr^{1}$-series) was measured in the low cliff above the platform.

Chapel Cove: Proterozoic and Placentian Series.—The Lower Cambrian nonconformably overlies the Proterozoic Holyrood granite (see discussion at Duffs locality). Erosional relief on the granite reaches 1.6 m, and the granite has spectacular fissure-fills up to 2.8 m deep that are filled with a calcareous to phosphatic, feldspathic quartz arenite with granite and phosphatic granules and pebbles. A 0–60 cm-thick black (phosphatic) conglomerate with granite, phosphatic, quartz, and dark grey and pink limestone clasts to 20x20x20 cm locally fills erosional depressions on the granite. This conglomerate is overlain by a burrowed, conoidal fossil-hash limestone

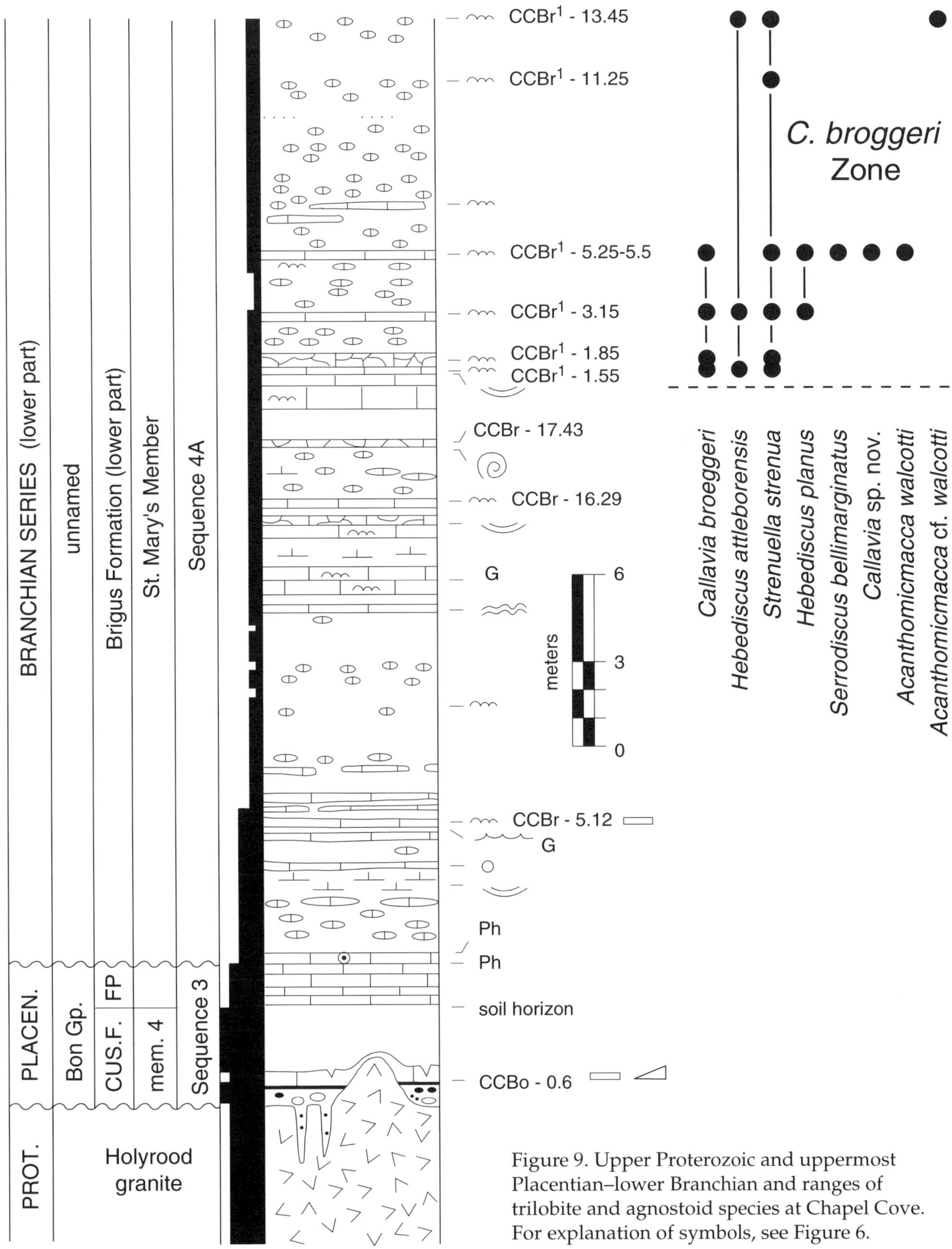

Figure 9. Upper Proterozoic and uppermost Placentian–lower Branchian and ranges of trilobite and agnostoid species at Chapel Cove. For explanation of symbols, see Figure 6.

(*Camenella baltica* Zone) with phosphatic laminae, lower planar cyanobacterial laminae, and upper burrow-mottled fabric. This limestone locally overlies knobs on the granite and has mud or synaeresis cracks on its top.

Strata immediately above the basal conglomerate and limestone at Chapel Cove are not present at Duffs. These include 2.3 m of recessive, black siliciclastic mudstone with an 11 cm-thick soil horizon at its top and 1.55 m of cliff-forming, dark grey, condensed, fossiliferous limestone (*Camenella baltica* Zone, note locally abundant *Gracilitheca bayonet* [Matthew] orthothecids) with a very low SH-V mat at its top. The very top of the cliff is a 20–30 cm-thick, brown-grey weathering, greenish lime mudstone with scattered oncoids, *Lingulella*, and phosphatic sand and granules. This latter limestone is separated from the underlying dark grey limestone by a phosphatic layer.

The dark grey siliciclastic mudstone and 1.55 m-thick limestone are member 4 of the Cuslett Formation and the Fosters Point Formation, respectively, of the upper Bonavista Group. The basal conglomerate and lowest limestone of the succession are the onlap deposits of member 4 of the Cuslett Formation. The presence of the upper Cuslett Formation and a thicker Fosters Point Formation than at Duffs are explainable by the closer location of the Chapel Cove section to the Placentia–Bonavista axis and earlier onlap and greater accumulation of the Bonavista Group at Chapel Cove. The Cuslett and Fosters Point Formations at Chapel Cove and Duffs illustrate facies unknown in other areas of southeastern Newfoundland or elsewhere in Avalonian North America. These units are dark grey to black and phosphatic in eastern and southern Conception Bay, and differ from the characteristic purplish to red colors of the non-phosphatic and presumably more unrestricted marine facies of these units elsewhere.

Chapel Cove: Branchian Series.—The 20–30 cm limestone at the top of the low cliff (described above) is identical to the basal limestone of the Brigus Formation at Duffs. This limestone forms the base of a 5.7 m, greenish siliciclastic mudstone and nodular to bedded limestone interval of the lower Brigus Formation that locally lacks trilobites. Trilobite remains appear almost precisely with a lithofacies change from green to red siliciclastic mudstones (Figure 9), but identifiable trilobites were only recovered from the highly indurated limestones of the section at the top of the massive (1.99 m) limestone that forms the vertical ledge on the west side of Chapel Cove Point and reappears to the northwest across the cove as an east-dipping platform. This massive limestone is the lateral equivalent of the 1 m-thick limestone in the upper Brigus Formation at Red Bridge Road (Figure 4).

Chapel Cove offers an excellent opportunity to examine a *Callavia broeggeri* Zone fauna from the St. Mary's Member of the Brigus Formation. Our collections have yielded a relatively diverse assemblage of at least eight species (Figure 9).

Highland Cove

Highland Cove: location.—The section at Highland Cove (Figure 10) is located just east of the Harbour Grace 1/50,000 Map Sheet, and lies south of Long Point and approximately 1.1 km northwest of Cavendish village (see Hutchinson, 1953, Map 1035A). It is reached by exiting Rte. 80 onto a dirt road that runs a short distance east to a cleared meadow that overlooks Trinity Bay. This intersection of Rte. 80 with the unimproved track is about 1.5 km north of Cavendish and about 0.5 km north of the town garbage dump. Final access to Highland Cove is by a series of hunters' and moose trails. The circuitous route leads southwest for about 1.5 km through a boggy terrane to high shore cliffs of black, pyritiferous siltstone-dominated MacLean Brook Group in the axis of the syncline that defines Highland Cove. The uppermost Chamberlain's Brook–lower MacLean Brook Group is best exposed in shore cliffs on the east limb of the syncline. The measured section begins at the high cliff (ca. 70 m) approximately 300 m south of the northeast head of Highland Cove.

Highland Cove: Acadian Series, Chamberlain's Brook and basal Manuels River Formations.—Little of the Chamberlain's Brook Formation is accessible on the east limb of the syncline at Highland Cove. The high cliff that forms the base of the measured section is a bedding surface 1.5 m below the top of

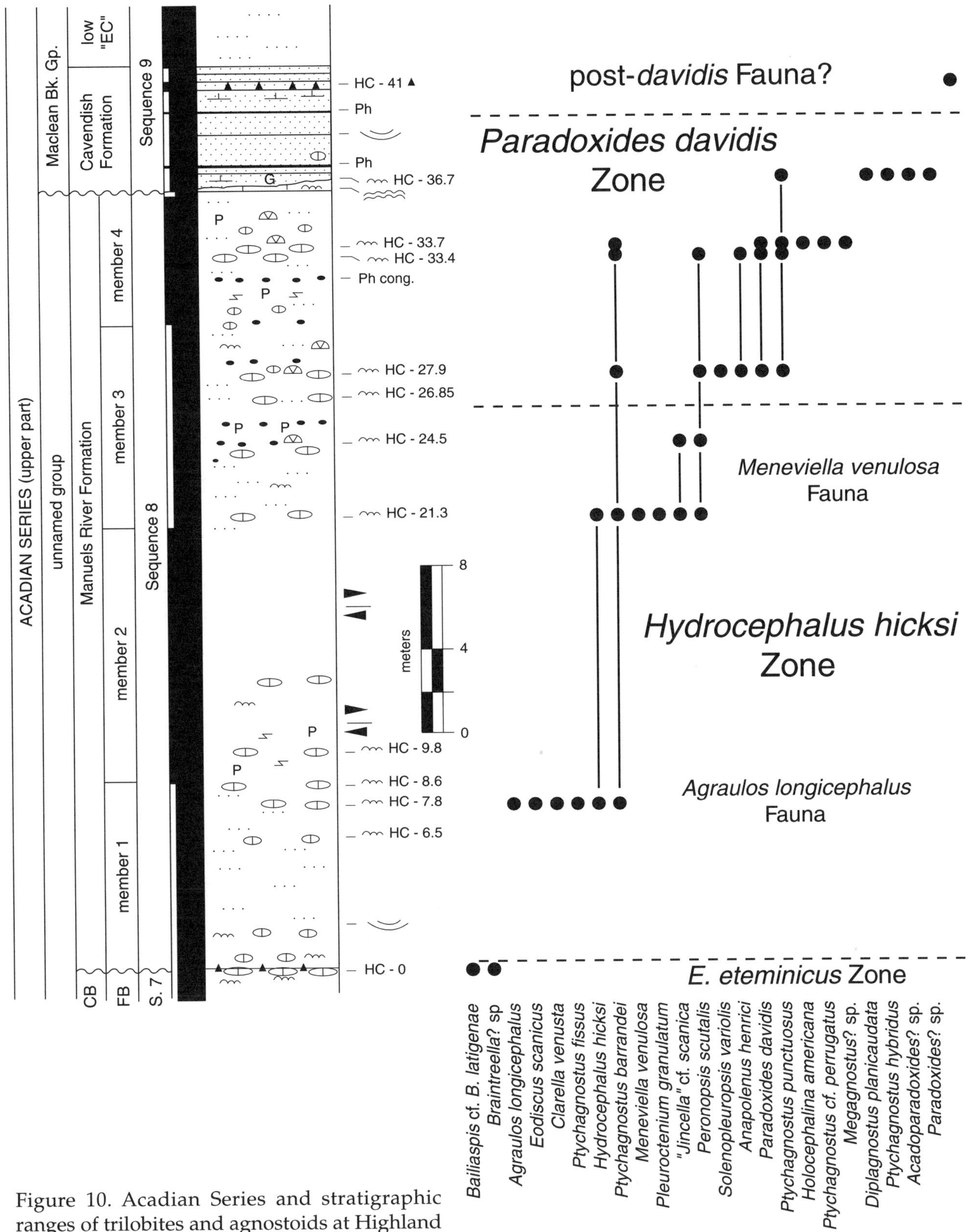

Figure 10. Acadian Series and stratigraphic ranges of trilobites and agnostoids at Highland Cove. For explanation of symbols, see Figure 6.

the Chamberlain's Brook Formation.

The contact of the Chamberlain's Brook and Manuels River Formations is marked by a deep, narrow (ca. 5 cm) reentrant filled with a calcite vein. Light grey, clayey "clots" occur locally in this fault calcite and represent the K-bentonite that is the base of the Manuels River Formation at other sections (August 12, Red Bridge Road and Manuels River; August 17, Fosters Point).

A bedding plane surface near sea-level with trilobites (e.g., *Hartella terranovica* Resser, *Badelusia tenera* [Hartt], and a paradoxidiid) characteristic of the Fossil Brook Member (depositional sequence 7) of the uppermost Chamberlain's Brook Formation occurs 0.5 m below the base of the Manuels River Formation. In addition, a thin (4 cm), nodular, white trilobite packstone is immediately below the basal Manuels River K-bentonite.

Highland Cove: Acadian Series, Manuels River Formation.—Highland Cove lies close to the St. Mary's–east Trinity axis (Figure 2) and has a thick section of the Manuels River Formation (36.4 m). This relatively great thickness (ca. 1.5 times that Manuels River and Fosters Point) is not only due to a somewhat greater rate of accumulation of compacted sediment, but also to the presence of an uppermost part of the formation (i.e., member 4, new, see Landing and Westrop, this volume) that is not seen at the Manuels River type section. A comparably thick Manuels River section is also present at the McLeod Point East section at the south end of Trinity Bay, a locality also on the St. Mary's–east Trinity axis (Figure 2, MPE; E. Landing and S.R. Westrop, unpublished data; see Hutchinson, 1962, p. 134, section 9). Most comments made on the Manuels River at Highland Cove are appropriate to the McLeod Point East section.

The lower 8.6 m of the Manuels River at Highland Cove consists of somewhat lighter grey siliciclastic mudstones, and is referred to member 1 of the formation. Abundant trilobites of the *Agraulos longicephalus* Fauna of the *Hydrocephalus hicksi* Zone occur in calcareous nodules 7.8 and 8.6 m from the base of the formation. The volcanic ash used to define the base of the darker grey and black siliciclastic mudstones with methanogenic nodules of member 2 at Manuels River and Fosters Point (August 12 and 17) is not present at Highland Cove, and the member 1–2 contact is defined at the color change from dark grey to black. This ash has probably been sheared out of the Highland Cove section, and is present at 11.9 m at McLeod Point East at the top of a somewhat thicker member 1.

Large, methanogenic, calcareous nodules (at 10.1 and 13.5 m) occur in lower member 2 and are absent in the upper, somewhat crumpled and bedding parallel-faulted black shales of member 2. A reappearance of somewhat lighter grey siliciclastic mudstones at 20.65 m and occurrence of articulated trilobites on large bedding plane surfaces in the interval 24.3–28 m, characterize member 3 at Manuels River and Fosters Point. The volcanic ash or bundle of thin ashes that occur near the base of member 3 at Manuels River and Fosters Point were apparently sheared out of the Highland Cove and McLeod Point East sections. Trilobites and agnostoids of the *Meneviella venulosa* Fauna of the *Hydrocephalus hicksi* Zone are in the lower half of member 3. Well-preserved elements of the *Paradoxides davidis* Zone fauna occur in nodules in the upper half of member 3 (Figure 10)

An important development seen only at two Manuels River Formation section that lie on the St. Mary's–east Trinity axis is the reappearance of black siliciclastic mudstones (at 30.2 m at Highland Cove) with numerous articulated *Paradoxides davidis* specimens (interval 32.25–35.1 m) at the top of the formation. Well-preserved, disarticulated, uncompacted material occurs in nodules in this interval (Figure 10). This 6.2 m-thick interval at Highland Cove is referred to member 4 of the Manuels River Formation. Member 4 comprises 5.5 m of very dark siliciclastic mudstone with articulated trilobites at the thickest known section of the Manuels River Formation (44.5 m) at Deep Cove, St. Mary's Bay (E. Landing and S.R. Westrop, unpublished data; see Fletcher, 1972).

Highland Cove: *Acadian Series, sequence 8–9 unconformity*.—Black shales of the upper Manuels River Formation are overlain by a 10–25 cm-thick, dark grey limestone with SH-V cyanobacterial build-ups, abundant black shale and phosphatic clasts, and deeply incised hypichnial traces (to 3 cm wide x 5 cm deep). The traces were produced on overcompacted mudstone and are comparable to those on the basal limestone of the older Fossil Brook Member of the Chamberlain's Brook Formation (August 12, Red Bridge Road). This limestone is laterally transitional into an intraclast pebble and cobble conglomerate. The upper 3 cm of shale immediately under the limestone is generally oxidized from black to light green.

Brownish-weathering, glauconitic, micaceous, somewhat calcareous, feldspathic sandstone with traces (*Planolites* and *Teichichnus*) dominates the 4.5 m of section above the stromatolitic limestone. These sandstones closely resemble the sandstones that overlie the conglomeratic sandstone on the Manuels River at Red Bridge Road (August 12), and are brought to the Cavendish Formation (new; Landing and Westrop, this volume). The top of the Cavendish Formation at Highland Cove lies at the abrupt contact with overlying laminated black shale and thin siltstones (at 42.15 m) and includes 30 cm of black volcaniclastic sandstone (40.85–41.15 m) that rests on the glauconitic sandstone and several higher beds of medium-grey siltstone and thin sandstone (41.15–42.15 m).

Oxidation of the shale at the top of the Manuels River Formation; presence of remanie clasts in a shallow-water, stromatolitic limestone; and occurrence of overlying glauconitic sandstones are developments that allow recognition of the depositional sequence 8–9 unconformity seen at other sections in southeastern Newfoundland (see Red Bridge Road, Manuels River, and Fosters Point) and elsewhere across North American and British Avalon (Landing, 1996a).

The stratigraphic assignment of strata at the top of the Manuels River Formation in this report follows Landing (1996a) and differs from Hutchinson (1953, 1962), who preferred to assign the interval from the base of the stromatolitic limestone almost to the top of the Cavendish Formation (new) to the Manuels River Formation. This assignment: 1) was not lithologically based and grouped strata significantly different from the typical fine-grained, dark grey to black lithologies of the Manuels River Formation into that unit; 2) did not account for the evidence for unconformity at the base of the limestone; and 3) was based on a biostratigraphic definition of the Manuels River to include strata to the top of the Middle Cambrian.

Highland Cove: biostratigraphy of sequence 8–9 unconformity.—New collections indicate that the sequence 8–9 unconformity at Highland Cove is in the uppermost Middle Cambrian, as at all other localities. Hutchinson (1962) interpreted the highest faunas at Highland Cove as the record of a post-*Paradoxides davidis* Zone interval that he referred to the *P. forchhammeri* Zone. Collections assigned to this interval by Hutchinson (1953, 1962) extended from the top 3 m of member 4 of the Manuels Rivers Formation into the overlying Cavendish Formation. Unequivocal cranidia and pygidia of *P. davidis* Salter are present in member 4 at HC-33.7, but poorly preserved paradoxidid material from the Cavendish Formation at HC-37.7 and HC-41 represent different species (Figure 10). However, agnostoid species in the basal beds of the Cavendish Formation include *Ptychagnostus punctuosus* (Salter), *P. hybridus* (Broegger), and *Diplagnostus planicaudata* (Angelin), all of which indicate a firm correlation with the *Ptychagnostus punctuosus* Zone (Westergaard, 1946; see Robison, 1984, for a discussion of the biostratigraphy of the *P. punctuosus* Zone), which is older than the basal *Goniagnostus nathorsti* Zone of the *P. forchhammeri* Stage. We agree with Rushton (1978, p. 249) that there is no evidence to support Poulsen and Anderson's (1975) belief that the upper Highland Cove section lies in the middle *Erratojincella brachymetopa* Zone of the *P. forchhammeri* Stage. Thus, the hiatus between the Manuels River Formation and the MacLean Brook Group at Highland Cove falls within the stratigraphic range of *Ptychagnostus punctuosus* and is unlikely to be lengthy. The top of the *Paradoxides davidis* Zone is not well constrained at Highland Cove, although a sample at 41 m yielded a single poorly preserved, deformed large-eyed species of *Paradoxides*? that is different from *P. davidis*.

Highland Cove: Acadian volcanism and sequence 8–9 unconformity.—As proposed by Landing (1996a), a number of sequence boundaries in the Cambrian of Avalon are temporally related to volcanism. This relationship suggests that they are explanable as the result of reorganizations in tectonic regime, and are not eustatically controlled. This relationship can be illustrated by developments at the sequence 8–9 contact at Highland Cove and other sections.

An association of volcanism with the sequence 8–9 unconformity is indicated by a black volcaniclastic sandstone (40.85–41.15 m) in the Cavendish Formation (5.7 m-thick) at Highland Cove. At the McLeod Point East section approximately 23 km to the south of Highland Cove (Figure 2, locality MPE), a slightly thicker section of Cavendish Formation (7.1 m) overlies the eroded top of member 3 of the Manuels River Formation, and has a 15 cm basalt pebble conglomerate at its base. Volcanic bombs up to 80x25 cm in cross-section occur in fine-grained sandstones in the middle of the Cavendish Formation at McLeod Point East. A massive cliff formed of 27 m of pillow basalts (Chapel Arm Member of McCartney, 1967) overlies the sandstones of the Cavendish Formation and are, in turn, overlain by black, micaceous siltstones of the lower MacLean Brook Group at McLeod Point East (E. Landing and S.R. Westrop, unpublished data).

Hutchinson (1962, p. 134) claimed to have collected late Middle Cambrian trilobites from thin limestone lenses in hollows between basalt pillows at the top of the Chapel Arm Member. However, his material remains undescribed, and our resampling of this horizon has yielded only unidentifiable fragments. If correctly identified, Hutchinson's (1962) data will confirm the temporal relationship of volcanism to the initial deposition of sequence 9 strata during the latest Middle Cambrian.

A slightly older episode of volcanism is associated with the end of sequence 8 deposition. Fletcher (1972) reported up to 60 m of calcitized pillow basalts (his "Hay Cove volcanics") in the upper Manuels River Formation in southern St. Mary's Bay. This strongly lenticular volcanic deposit thins southward to 30 cm at Deep Cove, where it is a thin calcitized ash bed in upper member 4, 2.5 m below the contact with the Cavendish Formation. The Hay Cove volcanics and Chapel Arm Member are not contemporaneous (fide Fletcher, 1972), but precede and follow development of the sequence 8–9 unconformity in southeastern Newfoundland (Figure 3).

August 14: Upper Proterozoic–Acadian of East Placentia Bay

Fortune Head

Fortune Head: location.—The Precambrian–Cambrian global stratotype section and point (see Landing, 1994) at Fortune Head lies southeast of the city of Fortune. The section (Figure 11) is reached by driving 2.7 km south of Fortune on Rte. 220 and turning west onto a gravel road. This road is followed for 3.3 km west across Fortune Brook and the high ridge. Vehicles can be parked near the plaque that marks Fortune Head as the Precambrian–Cambrian boundary global stratotype. The remainder of the route is by foot for several hundred meters north past the old town dump, with a final descent down a steep grassy slope to the shore (see Narbonne et al., 1987). The section dips northwest (15°), with dip increasing up-section. A completely exposed, very thick section (475 m) through upper member 2B of the Chapel Island Formation can be examined for one kilometer along the coast (Landing et al., 1988, p. 35–41). However, only about 40 m of strata through the Precambrian– Cambrian boundary interval will be examined. The significance of this stop for a Cambrian chronostratigraphy working group is that the base of the Cambrian is the base of the lowest Cambrian Placentian Series (Landing et al., 1989). Fortune Head is an Ecologic Preserve, and no collecting is allowed.

Fortune Head: Precambrian–Cambrian boundary and Placentian Series base.—Strata on the east side of the small cove at the Precambrian–Cambrian boundary stratotype locality and in the lower part of the low cliff on the west side of the cove belong to upper member 1 of the lower Chapel Island Formation. The 18.1 m of upper member 1 include: 1) lower red and green sandstones and shales with desiccation cracks and associated sedimentary features that indicate

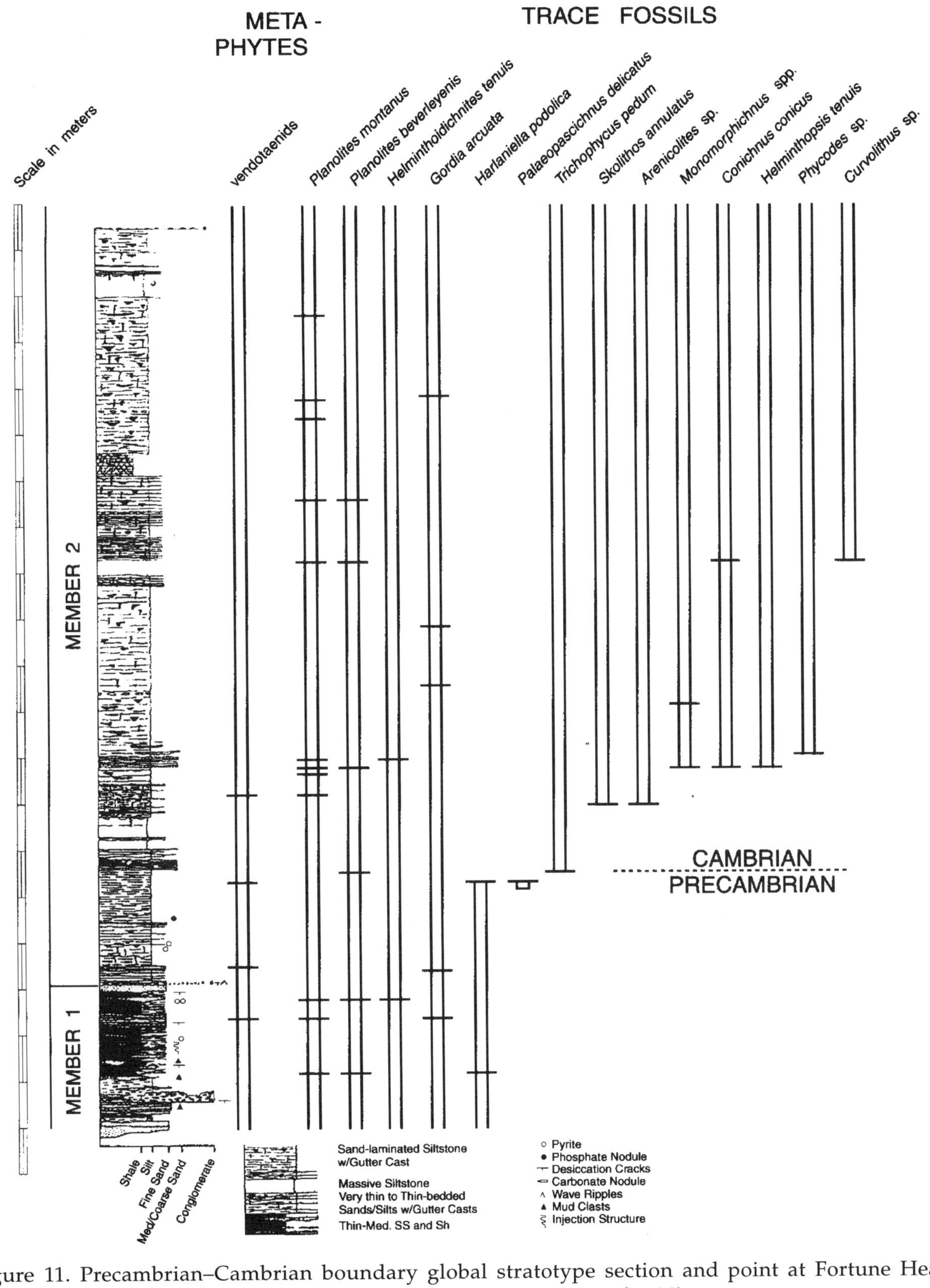

Figure 11. Precambrian–Cambrian boundary global stratotype section and point at Fortune Head. Modified from Narbonne et al. (1987, fig. 7) and Landing et al. (1988, fig. 20).

deposition in tidally influenced, peritidal conditions; and 2) upper, dark grey–black, laminated siltstones and shales with thin- to medium-bedded, grey-green sandstones with abundant synaeresis and few desiccation cracks that represent a restricted, shallow subtidal setting with intervals of subaerial exposure (Landing et al., 1988; Myrow et al., 1988).

The fauna of upper member 1 includes simple traces (*Planolites montanus* Richter, *P. beverleyensis* [Billings], *Harlaniella podolica* Fedonkin, and *Helminthoidichnites tenuis* Fitch [earlier reported as *Gordia marina* Emmons by Narbonne et al., 1987]) with body fossils in the darkest shaly units (*Sabellidites cambriensis* Yanishevsky, the vendotaenid *Tyrasotaenia* Gnilovskaya, and an impression of a soft-bodied "medusoid" organism). The traces, as those elsewhere in upper member 1 of the Chapel Island Formation, indicate the *Harlaniella podolica* Zone.

Dark grey–black facies of uppermost member 1 continue into the base of the low cliff on the west side of the little cove, where they are conformably overlain by member 2A. Green siltstones and thin to very thin sandstones of member 2A show gutter and pot casts with sandstone dikes and siltflow/liquified flow deposits. Facies and sedimentary structures of member 2A indicate rapid deposition on a storm-influenced, muddy delta front.

Facies changes from member 1 to the subtidal deposits of member 2A are not accompanied by faunal changes. *Tyrasotaenia* and the low diversity, subhorizontal traces of the *Harlaniella podolica* Zone (e.g., *Planolites*, *Gordia*, *Harlaniella podolica*, and *Palaeopascichnus*) persist into member 2A, with the disappearance of the latter two forms at 2.2 m above member 2A's base.

The fundamental behavioral and morphological changes that characterize the appearance of deep-burrowing and sediment-churning organisms at the beginning of the Cambrian (e.g., Narbonne et al., 1987) are marked by the appearance of the gallery-burrows of *Trichophycus pedum* 2.4 m above the base of member 2A (i.e., *Phycodes pedum* of earlier reports; see Geyer and Uchman, 1995). The lowest appearances of sedentary, burrow-dwelling organisms (*Skolithos* Haldeman and *Arenicolites* Salter), arthropod traces (*Monomorphichnus* Crimes), cnidarian burrows (*Conichnus* Myannil), and sediment-surface grazers with an elaborate browsing program (*Helminthopsis* Heer and *Gyrolithes*? de Saporta) also appear just above the lowest *Trichophycus pedum* Zone, and the trace fossil assemblage rapidly becomes Cambrian and Phanerozoic in character. No associated change in body fossils takes place through the Precambrian–Cambrian boundary GSSP at Fortune Head, and *Tyrasotaenia*, *Sabellidites*, and the chondrophorine *Kullingia delicata* (Fedonkin) persist into the Lower Cambrian. Specimens of the chondrophorine occur 13 m above the GSSP (Narbonne et al., 1991).

Little Dantzig Cove

Little Dantzig Cove: location.—The Little Dantzig Cove and Little Dantzig Cove Brook sections are reached by driving 23 km south of Fortune, Newfoundland, on Rte. 220 and turning east on the dirt road to Pieduck Point. Vehicles can be parked near the wharf, and participants will hike coastal paths for about 4.5 km to the middle part of the Little Dantzig Cove section (Figure 12; see Landing et al., 1988, figs. 24–26).

Pieduck Point is composed of resistant sandstones of the Random Formation, and overlying, steeply dipping beds of the Brigus and Chamberlain's Brook Formations are exposed at low tide in a southeast-striking syncline immediately north of the point. This Random–Chamberlain's Brook sequence is best exposed in a second syncline about 0.5 km north of Pieduck Point at the mouth of Snook's Brook. A section has been measured (by EL) that is very similar to that at Little Dantzig Cove Brook (Figure 13). The terrane north of the Snooks Brook syncline has outcrops of the Random Formation to the south side of Little Dantzig Cove. The east shore of Little Dantzig Cove is a high bluff of Pleistocene sediments, and a steeply southeast-dipping Random–Chamberlain's Brook section forms the northeast corner of Little Dantzig Cove at the outlet of Little Dantzig Cove Brook (Figure 12). The limestone-bearing middle part of the Chapel Island

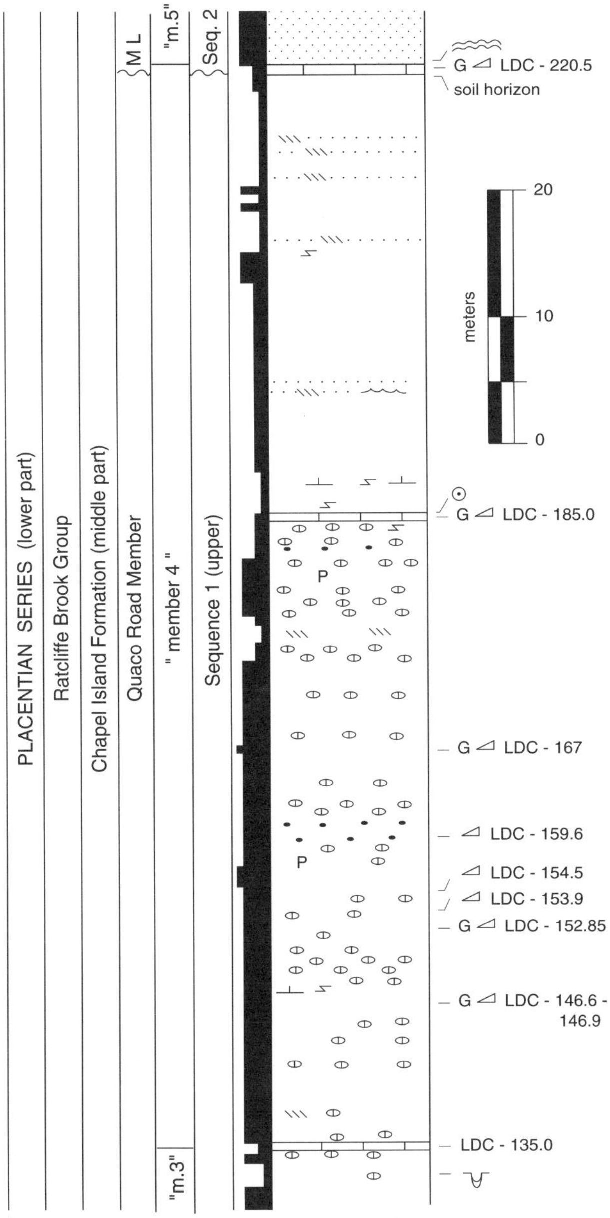

Figure 12. Middle Chapel Island Formation at Little Dantzig Cove. Figure shows informal members 3–5 of the formation of Bengtson and Fletcher (1983) as modified by Landing et al. (1988) and applies the sequence stratigraphic-based members (Quaco Road and Mystery Lake Members) of Chapel Island Formation in southern New Brunswick (Landing, 1996a). Abbreviation: ML, Mystery Lake Member. For explanation of symbols, see Figure 6. Figure modified from Landing et al. (1988, fig. 26; 1989, fig. 1).

Formation lies approximately 0.9 km northwest of the cliffs at the top of the Random Formation.

Little Dantzig Cove: middle Placentian Series and sequence 1–2 contact.—The lowest diverse small shelly fossils in southeast Newfoundland are associated with facies that indicate shoaling in the upper part of depositional sequence 1 in the Chapel Island Formation (see Landing et al., 1989, and summary above in "Biostratigraphic and geochronologic overview"). Rare specimens of the polychaete(?) *"Ladatheca" cylindrica* (Grabau) occur in the monospecific *"L." cylindrica* Zone in the deepest facies (greenish, "silvery" siliciclastic mudstones and siltstones) in the middle Chapel Island Formation, and are the oldest known skeletal remains in southeast Newfoundland. With shoaling higher in the section, a multispecies assemblage with snails (*Aldanella*) and rostroconchs (*Watsonella*), helcionellaceans (*Helcionella?*) appears in the green siliciclastic mudstones and nodular limestones of upper sequence 1. These taxa become very abundant in a 20 cm-thick interval of red, condensed peritidal limestones with cyanobacterial oncoids (Figure 12, horizon LDC-185), where they are joined by additional helcionellaceans (*Archaeospira?*, *Bemella?*), halkierids, pseudoconodonts (*Fomitchella*, *Mandeotaia*), hyoliths (*"Allatheca"*), and aragonitic problematica (*Anabarites*, *Coleoloides*).

These *Watsonella crosbyi* Zone taxa reappear in a condensed, thick, red limestone (Figure 12, horizon LDC-220) that has a 15 cm-thick soil horizon at its base. This soil horizon includes: 1) a lower reddish, mottled mosaic of pseudospar elongated parallel to bedding and argillaceous clots (i.e., an agglomeratic caliche fabric); and 2) an upper oncolitic packstone that is cut by an irregular to concavo-triangular crystalaria fabric and contains angular limestone intraclasts and fragments of Precambrian rhyolite (Landing et al., 1989). This soil horizon developed on the top of depositional sequence 1 prior to the marine onlap of depositional sequence 2 recorded by the upper part of the limestone (Landing et al., 1989; Myrow and Hiscott, 1993; Landing, 1996a). The upward lithofacies succession in the limestone includes: 1) lower sparse fossil wackestone with internal sediment- and calcite cement-filled sheet cracks; 2) a

"Ladatheca"-"Allatheca" wacke- to packstone with birdseye structures that is transitional laterally into broad, cyanobacterially-bound fossil wackestone domes with an Fe-Mn-impregnated, planar cyanobacterial cap; and 3) an *"Allatheca"-"Ladatheca"* pack- to wackestone that marks the reestablishment of less restricted near-shore conditions. The lowest occurrences of the pseudoconodont *Protohertzina anabarica* Missarzhevsky, helcionellacean *Anabarella plana* Vostokova, and tommotiids *Lapworthella ludvigseni* Landing and *Eccentrotheca kanesia* Landing, Nowlan, and Fletcher are at the top of the bed.

The persistence of *Watsonella crosbyi* Zone taxa through this stratigraphic break between depositional sequence 1 and 2 suggests, but does not prove, that the hiatus was short. This stratigraphic break within the Chapel Island Formation of southeast Newfoundland seems to be correlative with the type 1 sequence boundary in the Chapel Island Formation of southern New Brunswick (August 19, Hanford Brook East). For this reason, the designations "Quaco Road Member" and "Mystery Lake Member" that are applied, respectively, to the parts of the Chapel Island Formation below and above the sequence boundary in southern New Brunswick are applied in southeastern Newfoundland (Figure 12; see Landing, 1996a).

Peritidal environments disappear above the LDC-220 limestone that forms the base of sequence 2, and green to dark grey, thin- to medium-bedded sandstones and siltstones with parallel lamination and higher hummocky cross-stratified sandstones without any known body fossils characterize the upper Chapel Island Formation. These high energy, storm wave-dominated facies undergo a transition to tidal current-deposited sandstones of the Random Formation at the top of depositional sequence 2.

Little Dantzig Cove Brook

Little Dantzig Cove Brook: location.—As noted in the discussion of access to Little Dantzig Cove, the Little Dantzig Cove Brook section (Figure 13) is about 3.6 km northwest of Pieduck Point. The section is best exposed at low tide after storms have swept away the cobble beach.

Little Danzig Cove Brook: middle Placentian–Branchian Series unconformity and Brigus Formation.—Although only about 15 cm of relief is present on the Random Formation, this is a major unconformity. The upper part of the Placentian Series, or Bonavista Group (i.e., depositional sequence 2) is absent, and the Random Formation (top of depositional sequence 2) is overlain by the lower part of the Brigus Formation (depositional sequence 4A).

The Brigus Formation is only 12.3 m thick and includes: 1) a lower 30 cm of remanie sediment (calcareous quartz arenite reworked from the Random Formation with intraclast pebbles and a local Fe-Mn-impregnated SH-V cap); 2) 1.5 m of condensed, fossil (largely hyolith) wacke- to grainstones with numerous hardgrounds and abundant, low SH-V cyanobacterial build-ups; and 3) higher red and purple shale with bedded and nodular fossil wackestones with low SH-V build-ups, *Coleoloides*, and unidentifiable trilobite fragments.

Although the lower 1.5 m-thick limestone ledge is the lower part of the Brigus Formation, trilobites are exceptionally rare in the limestones, and are recognizable only as rare fragments. The fauna is dominated by hyoliths, *Coleoloides typicalis*, and small shelly fossils (e.g., the tommotiids *Camenella baltica, Eccentrotheca kanesia, Sunnaginia imbricata*) that persist from the sub-trilobitic Lower Cambrian. Relatively abundant valves of *Lingulella*, a brachiopod that does not occur below the Branchian Series, are present in acid residues.

Hutchinson (1962, locality 21904) noted trilobites from this limestone but did not further discuss them. We have not yet examined Hutchinson's collection, but believe that Fletcher (1972) is correct in assigning this Brigus Formation section to the lower part (i.e., *Callavia broeggeri* Zone) of the Brigus Formation.

The significance of the lower Brigus Formation at Little Dantzig Cove Brook is that the overall aspect of its fauna is "Tommotian" due to the extreme rarity of trilobites. This section and the trilobite-free lower part of the Hanford Brook Formation in southern New Brunswick (August 19, Hanford Brook West) negate Sepkoski's (1991)

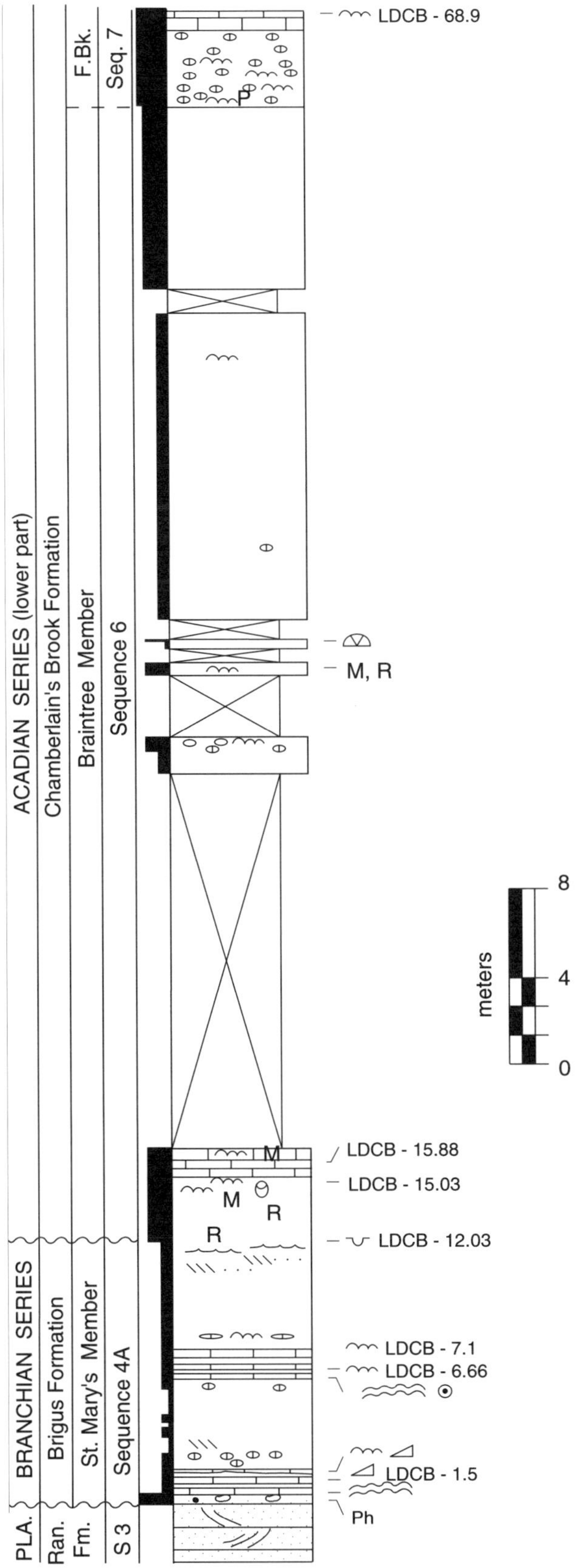

← Figure 13. Middle Placentian and Branchian–Acadian at Little Dantzig Cove Brook. Abbreviation: F. Bk., Fossil Brook Member. For explanation of symbols, see Figure 6. Modified from Landing et al. (1988, fig. 26).

proposal of a "retreat" of trilobites from near-shore paleocommunities in the Paleozoic. It would appear more likely that trilobites originated in and remained as a more off-shore component of Paleozoic communities through the Paleozoic (Westrop et al., 1995).

Little Dantzig Cove Brook: Acadian Series, Chamberlain's Brook Formation.—The contact of the Brigus and Chamberlain's Brook Formations is cryptic, and lies 12.3 m above the top of the Random Formation at the change from red siliciclastic mudstones with bidirectional (wave) microcross-beds into brown-grey (manganiferous) shales with rhodochrosite nodules and *Arenicolites* burrows. Although subtle, this unconformity represents a long hiatus between the lower part of the Brigus Formation (St. Mary's Member, or depositional sequence 4A) and, apparently, the middle part of the Chamberlain's Brook Formation (depositional sequence 6).

This correlation of the Chamberlain's Brook Formation is based on trilobites near the top of the section. Identifiable trilobites are rare in this thin section of the Chamberlain's Brook Formation, but include diagnostic forms at the top of the section near the outlet of Little Dantzig Cove Brook. Hutchinson's (1962) collection 21940, just below the massive 2 m-thick, black, manganiferous limestone at the top of the section, includes *Hartella terranovica* Resser and is assigned to the *Eccaparadoxides eteminicus* Zone. The presence of this bedded limestone in the usually limestone-free Chamberlain's Brook Formation and occurrence of this zonally important species allows reference of the top of the section to the Fossil Brook Member and depositional sequence 7. All of the lower part of the Chamberlain's Brook Formation is best refered to depositional sequence 6 and the Braintree Member.

August 15: Uppermost Placentian–Lowest Acadian, Western St. Mary's Bay

Branch Cove section

Branch Cove: location.—The important upper Branchian–lowest Acadian section at Branch Cove (Figure 14) was measured along the east limb of the gentle, north-south striking anticline that is exposed in sea cliffs 2 km east of Branch village. The section is reached by driving Rte. 100 north, then east of Branch village, and then parking several hundred meters south of the wooded hill crest. The core of the Branch anticline is reached by walking east through the pastures to the coast. Because of the gentle dip in the lower part of the section, the slaty Branch Cove section (Figure 14) extends for about a kilometer along the coast. To conserve time, the field trip visit will focus: 1) on volcanic ashes in the lower part of the exposed Brigus Formation (13.5–34.75 m); 2) on the Lower–Middle Cambrian (Branchian–Acadian Series) boundary; 3) on the appearance of a lowest Middle Cambrian ash (114 m); and 4) the lowest occurrence of *Acadoparadoxides*.

Branch Cove: upper Branchian Series.—The Brigus Formation at Branch Cove is dominated by red siliciclastic mudstones with a brownish cast. This is a representative section of the upper member (Jigging Cove Member [new]; Landing and Westrop, this volume). Lensing "splotches" of purple and green mudstone with a purple margin that is transitional into the surrounding red mudstone are common through the section. These mudstones are weakly calcareous, and a few poorly defined, incipient nodules that developed on looping horizontal traces occur in the red mudstones. Bidirectional (wave) microcross-beds and a few intervals with large-scale trough cross-beds of mud dunes are preserved in parts of the section that were not burrow-churned. Ten K-bentonite horizons are present; the ash at 27 m has been sampled for U-Pb zircon dating.

Fletcher's (1972) unpublished collections that were the basis for his uppermost Lower Cambrian trilobite zonation of southeastern Newfoundland were made at Branch Cove. These collections can be related to our measured section (Figure 14), although it should be noted that his measured thickness of the Brigus Formation at Branch Cove (95.7 m) is somewhat less than our measured thickness (103.05 m).

Fletcher (1972) assigned the lower 22.5 m of the Branch Cove section to an "*Acidiscus theristes* Zone", and the interval 22.5–76.5 m to a "*Bathydiscus* Zone" (see use of these zone in Bengtson and Fletcher, 1983). The definition of these "zones" is problematic (e.g., Landing, 1992). The lower "zone" includes *A. theristes* Rushton (5.1 m, 14.4 m, 15.6 m, 17.4 m) with the problematic species *Catadoxides magnificus* (14.4 m; see comments on *C. magnificus*, August 12, Manuels River bridge); an unillustrated *Condylopyge* (5.1 m); *Serrodiscus occipitalis* (Rasetti) (17.4 m); unidentified serrodiscines (5.1 m, 17.4 m) and pagetiidids (5.1 m); and ellipsocephaliids (17.4 m). A new *Bathydiscus* species that was the basis for the "*Bathydiscus* Zone" occurs below the "*Bathydiscus* Zone" at 15.6 and 17.4 m), and it is evident from Fletcher's (1972) data that *Bathydiscus* does not distinguish the "*Acidiscus theristes*" and "*Bathydiscus*" Zones.

Most of Fletcher's (1972) second and middle trilobite zone at Branch Cove, the "*Bathydiscus* Zone", is a barren interval. Three closely spaced horizons in the middle of this "zone" have trilobites. These trilobites include the new *Bathydiscus* species (45 m); *Serrodiscus ctenoa* Rushton (41 m, 44 m, 45 m); *S. occipitalis* (45 m); and unidentified forms of an olenellid, conocoryphid, redlichiine, and, surprisingly, a paradoxidiid (all at 45 m). The only trilobite seemingly unique to the "*Bathydiscus* Zone" is *S. ctenoa*, although absence of this species lower in the Branch Cove section may be due to collecting bias and the small number of specimens that can be collected from the mudstones.

The third and uppermost zone named by Fletcher (1972, also Bengtson and Fletcher, 1983) in the Brigus Formation at Branch Cove is the *Oryctocara granulata* Zone (76.5 m to top of formation). This zone is based on two collections. The lower collection at 78.6 m (with *Serrodiscus occipitalis*, the new *Bathydiscus* species, and reported paradoxidid fragments) is indistinguishable on

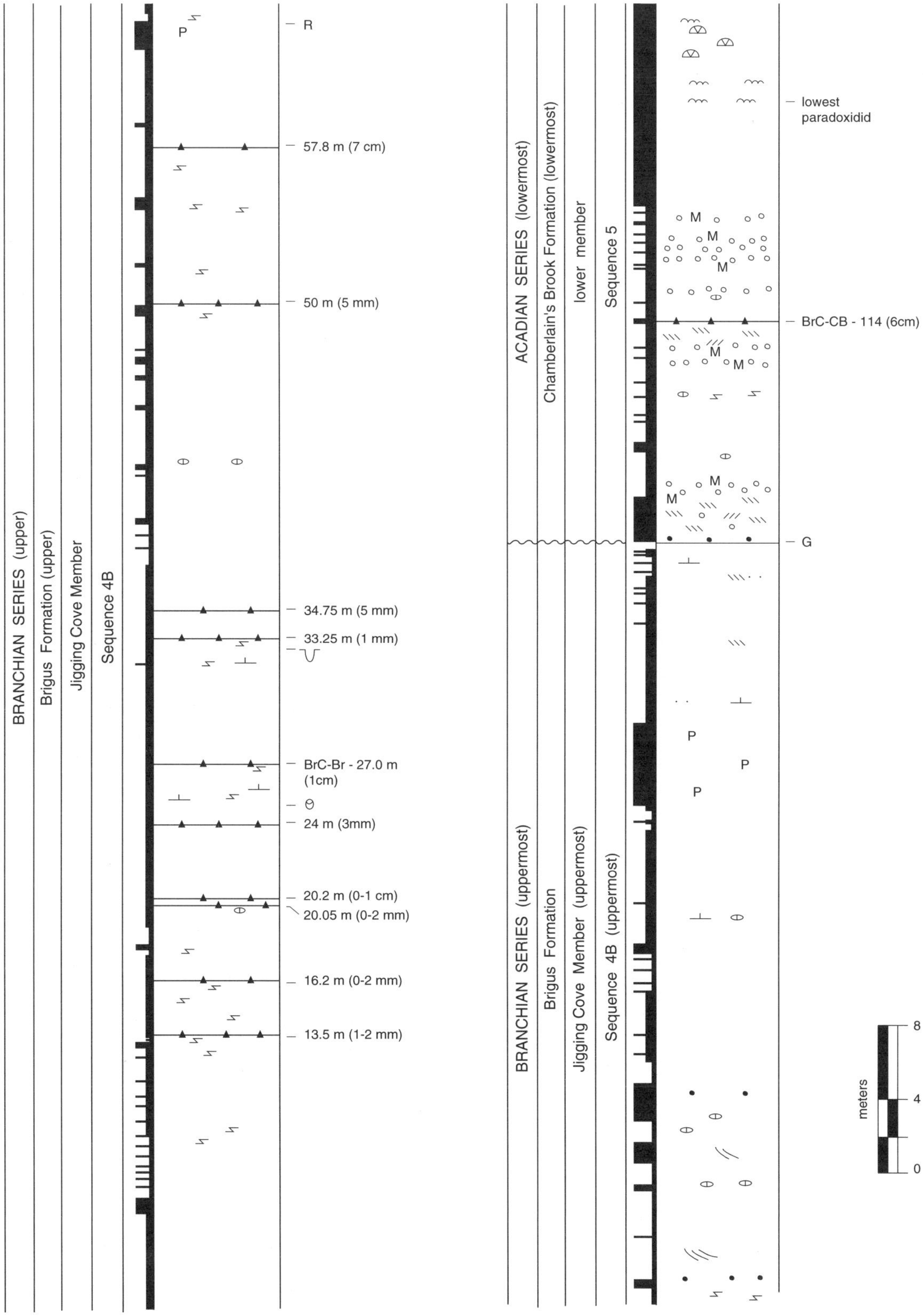

Figure 14. Upper Branchian–lowest Acadian on north side of Branch Cove, eastern St. Mary's Bay. For explanation of symbols, see Figure 6.

the basis of species recovered from collections in Fletcher's (1972; Bengtson and Fletcher, 1983) "*Bathydiscus* Zone". Only the upper collection, taken about 1 m below the contact with the Chamberlain's Brook Formation (with *Oryctocara granulata* Tchernysheva, *Acidiscus* sp. cf. *A. theristes*, the new *Bathydiscus* species, *B.* sp. cf. *B. dolichometopus* Rasetti, a *Condylopyge*, *Serrodiscus occipitalis*, *S.* sp. *S. latus* Rasetti, undescribed *Peronopsis* spp.), can be referred to an *Oryctocara granulata* Zone on the basis of its fauna. As noted by Robison et al. (1975), *O. granulata* was first described from the Amgan Stage, and this fauna from the top of the Brigus Formation is Middle Cambrian in aspect by comparison with eastern Siberian sections.

Branch Cove: lower Acadian Series.—The base of the Chamberlain's Brook Formation is a cryptic unconformity between depositional sequence 4B (Jigging Cove Member of the Brigus Formation) and a lower, unnamed member (depositional sequence 5) of the Chamberlain's Brook Formation at Branch Cove (Landing, 1996a; Figure 3). The lower 2.15 m of the Chamberlain's Brook Formation is composed of prominently microcross-bedded, green grey, fine-grained sandstone with limonitic nodules that rests on the primarily reddish siliciclastic mudstones of the uppermost Brigus Formation. The 15 cm of siliciclastic mudstone at the top of the Brigus Formation is bleached to a light brown color not found elsewhere in the section, and is overlain by a horizon with scattered large quartz grains, phosphate granules, and sparse manganese nodules at the base of the Chamberlain's Brook Formation.

Trilobites reported by Fletcher (1972) from the lower Chamberlain's Formation at Branch Cove are lowest Middle Cambrian and have been divided into two somewhat confusing biostratigraphic subdivisions. The lowest collections include undescribed new species of *Condylopyge* (3.3 m) and *Peronopsis* (3.3 m, 16.5–20.1 m) that persist from the underlying Brigus Formation; a new *Pagetides* species (3.3 m, 9.7 m, 16.5–20.1 m); a form similar or identical to *Braintreella rogersi* (Walcott) (3.3 m, 6.6 m, 16.5–20.1 m); *Acadoparadoxides harlani* (Green) (16.5–20.1 m); and *Skreiaspis quadrangularis* (Whitfield) (16.5–20.1 m).

The lower 20.1 m of the Chamberlain's Brook Formation at Branch Cove have been referred to a "*Pagetides* Local-Range biozone". The overlying strata (ca. 20.1–57 m), which incidently have the same species as the "*Pagetides* Local-Range biozone" but lack the new *Pagetides* species, were referred to an "*Acadoparadoxides harlani* Acme-Biozone" (Fletcher, 1972; Bengtson and Fletcher, 1983). It is our impression that the abrupt appearance of *Acadoparadoxides harlani*, at 22 m above the base of the Chamberlain's Brook Formation in our measured section, is a convenient biostratigraphic marker within this section. We collected a 6 cm-thick K-bentonite 11 m above the base of the Chamberlain's Brook Formation and 11 m below the lowest occurrence of *A. harlani.* This ash may provide a U-Pb zircon date on the lowest Middle Cambrian.

Jigging Cove

Jigging Cove: location.—A completely exposed, east-dipping Lower Cambrian section lies on the northwest shore of Jigging Cove (Figure 15). The section is 10 km northeast of Branch village and is reached from Rte. 100. A rough road just south of the bridge across Red Head River leads to the shore. The base of the section is reached by wading across Red Head River near the storm beach. The Jigging Cove section was measured north from the faulted outcrop of uppermost Random Formation (approximately 7 m exposed) in the axis of the anticline. It continues along the northwest shore of Jigging Cove into the faulted axis of the syncline at the north end of Jigging Cove. Hutchinson's (1962, p. 139, 140) Jigging Cove section was measured in the sea cliffs south of the outlet of Red Head River.

The focus of this stop will be on the unconformities at the Random–Fosters Point and Fosters Point–Brigus contacts near the base of the section. In addition, the type sections of the newly proposed St. Mary's and Jigging Cove Members of the Brigus Formation lie at Jigging Cove (Landing and Westrop, this volume). The regional unconformity between these two members is exposed in a small cove 0.75 km north of the mouth of Red Head River.

Jigging Cove: Placentian Series.—Medium- to massive-bedded quartz arenites of the Random Formation are overlain by 4.5 m of Fosters Point Formation (upper *Camenella baltica* Zone). The lower Bonavista Group succession of Petley, West Centre Cove, and Cuslett Formations that reach 275+ m on the Placentia–Bonavista axis (Landing and Benus, 1988b) are absent. This Random–Fosters Point (middle–latest Placentian Epoch) hiatus is identical to the hiatus between the Jees and Home Farm Members at Nuneaton, Warwickshire (see Brasier, 1984, 1989).

The lowest bed of the red limestone-dominated Fosters Point Formation is a 35–50 cm-thick unit of reworked Random Formation sand that forms a calcareous and limonitic quartz arenite to quartz granule conglomerate with shale clasts. Sparse fossil wackestone (*Coleoloides*-dominated) dominates the overlying 3.0 m of the Fosters Point Formation, and a regionally extensive *Coleoloides*-algal mud mound biostrome with in situ, vertical *Coleoloides* (e.g., August 16, Smith Point) in the upper Fosters Point Formation occurs at 3.4–3.55 m. The top 0.95 m of the Fosters Point Formation consists of fossil wacke- to packstone with limonite-impregnated oncoids, which are locally truncated on their upper surfaces or locally develop into low SH-V structures.

Jigging Cove: Branchian Series.—A 15 cm-thick, pinkish, trilobite wacke- to packstone abruptly overlies the Fosters Point Formation and locally overlies truncated large oncoids. This bed at the base of the Brigus Formation and sequence 4A marks the onlap of the Branchian Series across an unconformity on the sub-trilobitic Placentian Series in St. Mary's Bay.

The overlying lower part of the Brigus Formation is dominated by structureless (burrow-churned), red and purple (with green lenses), slaty, siliciclastic mudstone with thin (to 20 cm), nodular to bedded limestones that contain sparse fossil debris. Nodular limestones somewhat higher in the formation (47.7–63.4 m) lack significant trilobite sclerites and developed in more porous and permeable intervals in the purplish-red to red siliciclastic mudstones on large horizontal traces.

An abrupt transition from red- and purple-red-dominated siliciclastic mudstones into a 14 m-thick interval of pyritiferous, laminated, green mudstones takes place at about 87 m in the section. This abrupt transition is preceded by the appearance of trilobite-bearing, calcareous nodules in red, burrow-churned siliciclastic mudstones (84.7–86 m) and an exceptionally thick, reddish, trilobite wacke- to packstone (86–86.8 m).

Although the top of the thick trilobite-bearing bed is sheared, slabbed sections show that its top 10 cm have limonitic oncoids, small SH-V stromatolites, and an overlying, thin (0–3 mm) phosphatic cap on an eroded, undulatory upper surface. This surface is overlain by a calcareous, fine-grained, glauconitic (now chloritized), burrowed (*Teichichnus*) sandstone (86.8–87.3) with limonite pisolites, phosphatic pebbles, and rare Proterozoic volcanic pebbles. Trilobite-bearing nodules developed in its upper 30 cm. About 14 m of phosphate nodule-bearing, pyritiferous, laminated siliclastic mudstones with methanogenic calcareous nodules overlie the trilobite-bearing nodules.

The lithofacies changes in this part of the Brigus Formation reflect: 1) shoaling at the top of the lower member of the Brigus Formation into shallower facies associated with oncoid formation; 2) marine offlap that led to erosion and transport of reworked clasts derived from the basin (phosphatic clasts) and adjacent subaerial areas (Proterozoic volcanics); and 3) onlap that led to phosphate impregnation of the top of the lower member, deposition of glauconitic sandstone in normal marine environments, and continued deepening into dysaerobic (laminated green mudstone) facies. The phosphatized layer on the top of the pink, trilobite-bearing limestone marks the sequence 4A–4B contact within the Brigus Formation. The lower part of the Brigus Formation at Jigging Cove is the type section of the St. Mary's Member (new), and the sequence boundary marks the base of the type section of the Jigging Cove Member (new) of the Brigus Formation (Landing and Westrop, this volume).

Identifiable trilobites are rare, but demonstrate that *Callavia broeggeri* Zone faunas extend through at least a 50 m interval. The lowest

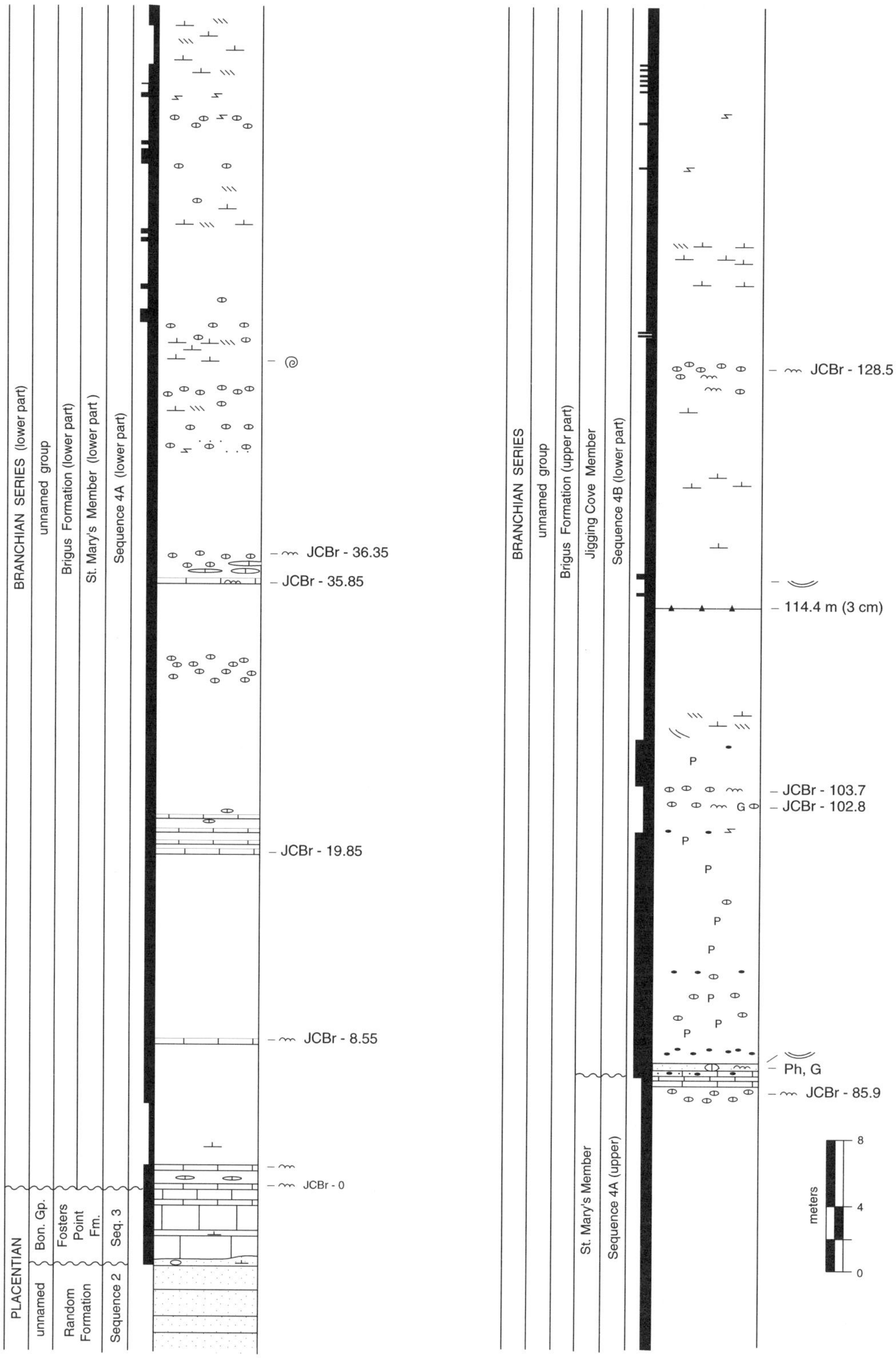

Figure 15. Middle and upper Placentian and Brachian at Jigging Cove, eastern St. Mary's Bay. Sequence has the type sections of the St. Mary's and Jigging Cove Members (new; Landing and Westrop, this volume). For explanation of symbols, see Figure 6.

appearance of identifiable specimens is at 37.35 m [*Strenuella strenua* (Billings), *Callavia* sp. nov., *Hebediscus planus* (Hutchison)], and their highest occurrence is at 85.9 m [*Strenuella strenua* (Billings), *Callavia broeggeri* (Walcott), *Serrodiscus bellimarginatus* (Shaler and Foerste)].

Above the green shale-dominated intervals of the lower Jigging Cove Member (87–100.9 m and 103.85–106.5 m), the dominant facies differ somewhat from the underlying St. Mary's Member. The Jigging Cove has a more brownish cast of the red siliciclastic mudstone, and lenticular (to 0.5 m-thick) green- to purplish-colored mudstone lenses appear in the St. Mary's Member. Such primary sedimentary structures as numerous burrows (particularly *Teichichnus*); numerous, thin but laterally continuous, whitish layers to 2 cm thick; and a few low-angle trough cross-sets (i.e., mud dunes) occur in the Jigging Cove Member, but are rare in the St. Mary's Member. The greater tendency for preservation of these primary sedimentary structures in the Jigging Cove Member suggests that burrowing intensity was reduced during deposition. A 3.0 cm-thick feldspathic ash appears in a small waterfall at 114.4 m. Very small calcareous nodules persist through the Jigging Cove Member, but only one of these (at 128.5 m) yielded a trilobite specimen (*Strenuella*? sp.).

August 16: Uppermost Placentian–Lower Acadian, Smithsound, Western Trinity Bay

Smith Point

Smith Point: location.—The classic Smith Point section (Walcott, 1900b) lies northeast of Clarenville on the north shore of Smith Sound. It is reached by driving east of Georges Brook on Rte. 232 and passing through the village of Harcourt. The east end of Harcourt is contiguous with the hamlet of Fosters Point (Jenness, 1963, Map 1130A West Sheet). A steep paved lane, approximately 150 m long, descends to Smith Point. The government wharf is constructed on the lowest Brigus Formation, and the low limestone ledges immediately to the east are the type locality of the "Smith Point Limestone" (designation abandoned; replaced by uppermost Placentian Fosters Point Formation and unconformably overlying, basal limestone of the Brigus Formation [Landing and Benus, 1988a], Figure 16).

Smith Point: Placentian Series.—An incomplete section of the middle–upper Bonavista Group is exposed east of Smith Point. A 124.15 m-thick section was measured through the red and purple shales with nodular and bedded limestones of the Cuslett Formation, with the top 7.6 m formation composed of the upper, limestone-poor, dominantly green shale of member 4 (Figure 16; Landing and Benus, 1988a). The top of member 4 is red, and limestones appear at its top.

The top of the Bonavista Group is composed of the 7.0 m-thick type section of the limestone-dominated Fosters Point Formation. The base of the Fosters Point Formation is drawn at the level that the limestone beds become thicker than the intervening shales in this depositionally unbroken Cuslett–Fosters Point section. The red color of the limestone here and at Jigging Cove (August 15) is characteristic of the Fosters Point Formation, and the grey, phosphatic limestone seen at Duffs and Chapel Cove (August 13) is limited to the near-shore facies at the south end of Conception Bay.

The nodular-appearing sparse fossil wackestones of the lower Fosters Point form a griotte with the red shale that surrounds the nodules. Most of these nodules are likely algal mud mounds with stylolitized contacts with the surrounding shale. Shell hash (primarily *Coleoloides* tubes) often fills depressions between the griotte nodules.

A coarse-grained shell hash lag with wave-oriented specimens of a diverse upper *Camenella baltica* Zone fauna (e.g., the large helcionellid *Randomia aurorae* Matthew, the snail *Aldanella attleborensis* (Shaler and Foerste), the orthothecids *Gracilitheca bayonet* (Matthew) and *Orthotheca degeeri* (Holm), and conchs and opercula of *Hyolithes tenuistriatus* Linnarsson s.l.) appears 6.4 m above the base of the Fosters Point Formation. This shell lag may have supplied the material for Matthew's (1899) report on the sub-trilobitic "Etcheminian" fauna from eastern Newfoundland. Well-defined mud mounds with in situ, vertical *Coleoloides* tubes that radiate to the margins of the mounds appear in the top bed of the Fosters Point Formation and form a "worm-reef".

Smith Point: Branchian Series.—The top of the Fosters Point Formation marks the Placentian–Branchian Series depositional sequence boundary across eastern Newfoundland. The top of the Fosters Point has ca. 15 cm of relief, and the oncolitic trilobite wackestones of the lowest Brigus Formation (108.6 m thick at Smith Point) with *Callavia broeggeri* Zone faunas (Hutchinson, 1962; *Callavia?* sp. and *Strenuella strenua?* have been found in our work) pinch out against topographic highs on this surface. The lower limestones of the Brigus Formation are 1.3 m in thickness and are somewhat lighter red in color than those of the Fosters Point Formation. The top of the Fosters Point Formation is slightly recrystallized and has a bluish color immediately below the Brigus Formation. The regular appearance of this "blue bed" (Landing and Benus, 1988b) may mark incipient soil development at this contact. It should be noted that the *Coleoloides*-algal mound "worm reef" lies immediately below the Fosters Point—Brigus unconformity at Smith Point. This same horizon is 1.0 m below the base of the Brigus Formation at Jigging Cove, and suggests deeper erosion and/or an earlier end to accumulation of Fosters Point Formation limestones at Smith Point (E. Landing, unpublished data).

The lower Brigus Formation above the basal limestone is covered. Structureless (burrow-churned) red, purple and green mudstones crop out east of the wharf and are capped by a resistant, 80 cm-thick ledge in the shore cliff. This ledge is succeeded by green pyritiferous shale, and consists of a lower, 20 cm-thick, pink nodular trilobite wackestone and an upper, 60 cm-thick, green, glauconitic, calcareous sandstone with limestone and phosphatic pebbles at its base. *Callavia broeggeri* Zone trilobites are not present above the nodular limestone, and the top of the calcareous sandstone has *Myopsostrenua?* sp. cf. *M.? sabulosa* (S.R. Westrop and E. Landing, unpublished data). These stratigraphic relationships and the change in trilobite faunas are identical to the depositional sequence 4A–4B and the St. Mary's–Jigging Cove contact at Jigging Cove. The Jigging Cove Member at Smith Point also shows the same brownish-red coloration with lenticular greenish and purple mudstones as the Jigging Cove Member at Jigging Cove. The thin volcanic ashes characteristic of the Jigging Cove Member at Branch Cove (August 15, Figure 14) are represented by a thin volcanic ash 3.1 m below the top of the Jigging Cove Member at Smith Point.

Smith Point: unconformable base of Acadian Series.—A 90 cm-thick, black-weathering rhodochrosite and manganese oxide nodule bed (after calcareous nodules) with scattered quartz granules defines the base of the Chamberlain's Brook Formation several hundred meters west of the Smith Point wharf. Matthew (1899) believed that this bed was a basal Cambrian conglomerate that overlay, and was the result of tectonism of, his sub-Cambrian and sub-trilobitic Etcheminian Series. Walcott (1900b) subsequently pointed out the presence of Lower Cambrian trilobites in the underlying strata now referred to the Brigus Formation, and used this data to argue against Matthew's (1889) concept of an Etcheminian Series.

To some extent, Matthew (1899) was correct in believing that this nodule bed reflects an unconformity in the section. The Jigging Cove Member (depositional sequence 4B) is overlain by a *Paradoxides bennetti* Zone fauna (depositional sequence 6) that occurs in the manganiferous basal bed of the Chamberlain's Brook Formation (Hutchinson, 1962, collection 20389).

A 2 cm-thick volcanic ash lies 4.4 m above the base of the Chamberlain's Brook Formation. This ash and the uppermost Brigus Formation ash at SPBr-105.75 (Figure 16) have been sampled to provide a bracket on the Lower–Middle Cambrian boundary in Avalon.

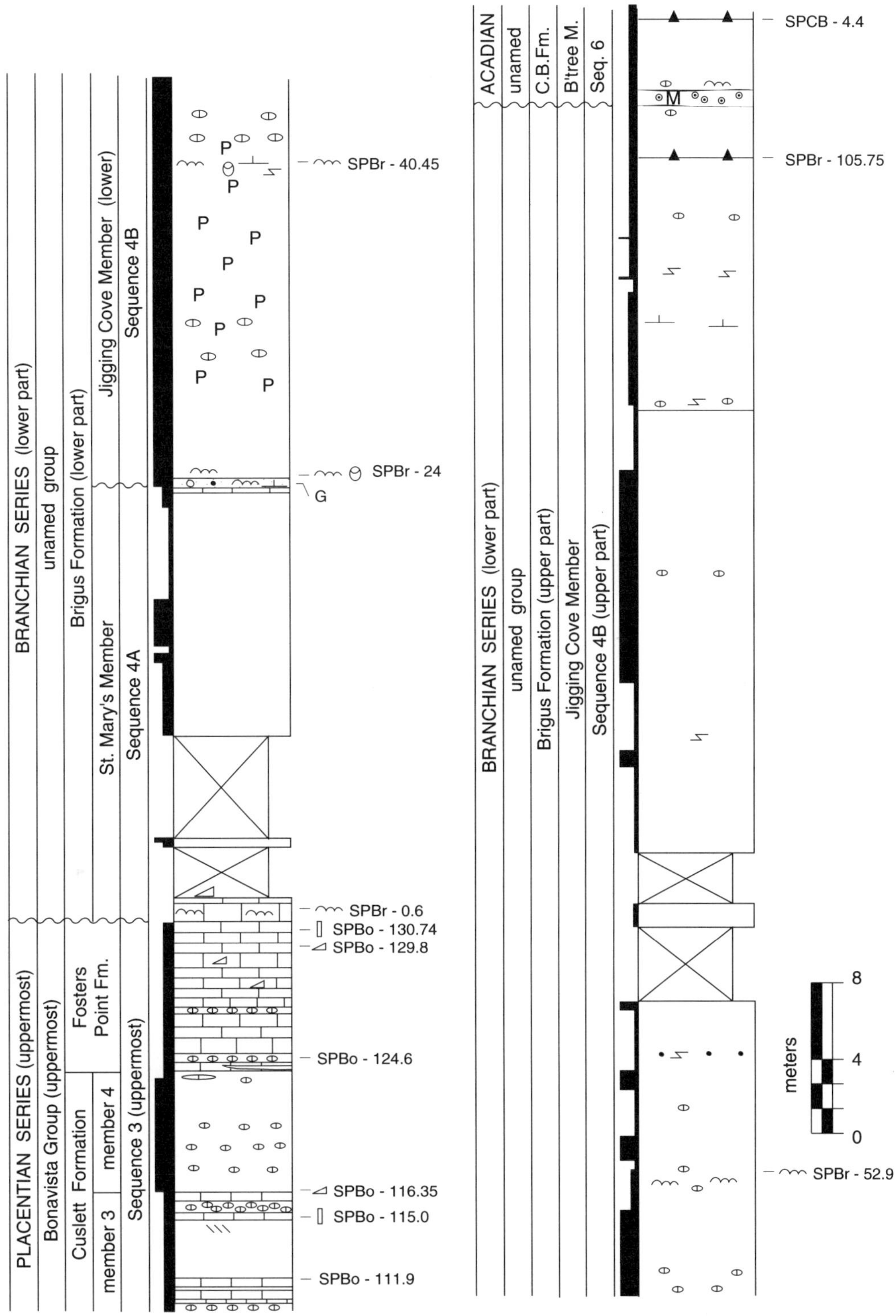

Figure 16. Upper Placentian–lower Acadian at Smith Point, western Trinity Bay, shows unconformable bases of Branchian and Acadian series. Lower part of Cuslett and upper Chamberlain's Brook Formations not shown. Abbreviations: B'tree M., Braintree Member; C.B. Fm., Chamberlain's Brook Formation. For explanation of symbols, see Figure 6. Detailed lithology of uppermost Cuslett-lowest Brigus Formations in Landing et al. (1988, fig. 31).

AUGUST 17: MIDDLE PLACENTIAN AND ACADIAN–MERIONETHIAN, WESTERN TRINITY BAY

Fosters Point

Fosters Point: location.—The designation "Fosters Point" is given to an unmetamorphosed, steeply west-dipping section approximately 1 km northwest of Fosters Point on the south side of Random Island. It is Hutchinson's (1963, p. 143, 144) Section 17 "south of Elliott's Cove and north of Fosters Point" and lies closest to Fosters Point. The section is reached by exiting Rte. 230 in the village of Milton onto Rte. 231. Rte. 231 is followed along the south shore of Random Island through Weybridge to Fosters Point, where parking is available at the wharf. The upper Chamberlain's Brook Formation–lower MacLean Brook Group sequence (Figure 17) forms a shore cliff that transects the east limb of a large syncline.

Fosters Point: structure and acritarchs from basalt(!).—Dark grey, upper Proterozoic sandstones exposed on the shore west of the wharf are overlain by Random Formation quartz arenites (middle Placentian). The Random Formation is faulted against red siliciclastic mudstones with lower nodular and higher medium- to massive-bedded limestones of member three of the upper Cuslett Formation (upper Placentian) that crop out where the shoreline bends north. The Cuslett Formation is fault-duplicated along the shore, and manganiferous, red siliciclastic mudstone of the lower Chamberlain's Brook Formation (lower Acadian) are faulted against the Cuslett Formation. Black amygdaloidal basaltic dikes that are likely Jurasic occur along this part of the shore. They were reported by martin and Dean (1981, 1988) as black sandstones. Martin and Dean sampled them and claimed to have recovered Cambrian acritarchs.

Fosters Point: Acadian Series, uppermost Chamberlain's Brook–Manuels River Formations.—The upper Chamberlain's Brook Formation is relatively undeformed, and a completely exposed section extends through the Manuels River Formation into the MacLean Brook Group. The detailed stratigraphy of this part of the Fosters Point section is comparable to that at Manuels River (August 12) and Highland Cove (August 13).

A thin (14 cm), fault-duplicated, trilobite packstone with large burrows at its base lies 2.4 m below the base of the Manuels River Formation and marks the base of the Fossil Brook Member of the uppermost Chamberlain's Brook Formation. The basal Manuels River K-bentonite is thick (ca. 10 cm) and forms a deep reentrant. As at Manuels River, the greyer shales of the lower part (member 1) of the Manuels River Formation terminate against a K-bentonite (at 5.69 m; 5.9 m at Manuels River) and are overlain by darker grey to black shales with methanogenic nodules of lower member 2. A third ash (17.64–17.66 m) at Fosters Point lies at about the same position as the upper ashes at Manuels River (interval 13.5–14.2 m) and just below horizons with articulated specimens of *Paradoxides davidis* (18.55–19.2 m at Fosters Point and 18.4 m at Manuels River. These articulated specimens disappear in the uppermost Manuels River Formation at Fosters Point and Manuels River. The base of member three at the top of the Manuels River Formation is defined slightly below the upper ash with the appearance of thin siltstones and fine-grained sandstone laminae in black shale at both sections.

Nodules in the lower part of member 2 (5.78–8.43 m) yield diverse (up to 10 species) well-preserved elements of the *Agraulos longicephalus* Fauna of the *Hydrocephalus hicksi* Zone. An interval of nodules in member 3 is characterized by *Paradoxides davidis* Zone trilobites and agnostoids that include *Ptychagnostus barrandei* (Hicks), *P. punctuosus* (Salter), *Anapolenus henrici* Salter, and *P. davidis* Salter.

Fosters Point: Acadian Series, sequence 8–9 boundary.—Stratigraphic relationships and biostratigraphy above the Manuels River Formation are comparable to those at Manuels River. A yellow weathering (pyritiferous), 5 cm-thick, fine-grained, conglomeratic sandstone with black shale and phosphatic clasts lies on the *Paradoxides davidis* Zone of the upper Manuels River Formation as the lowest bed of the MacLean Brook Group. As at Manuels River on the east side of Conception Bay, this thin conglomeratic sandstone on the west side of Trinity Bay is the feather

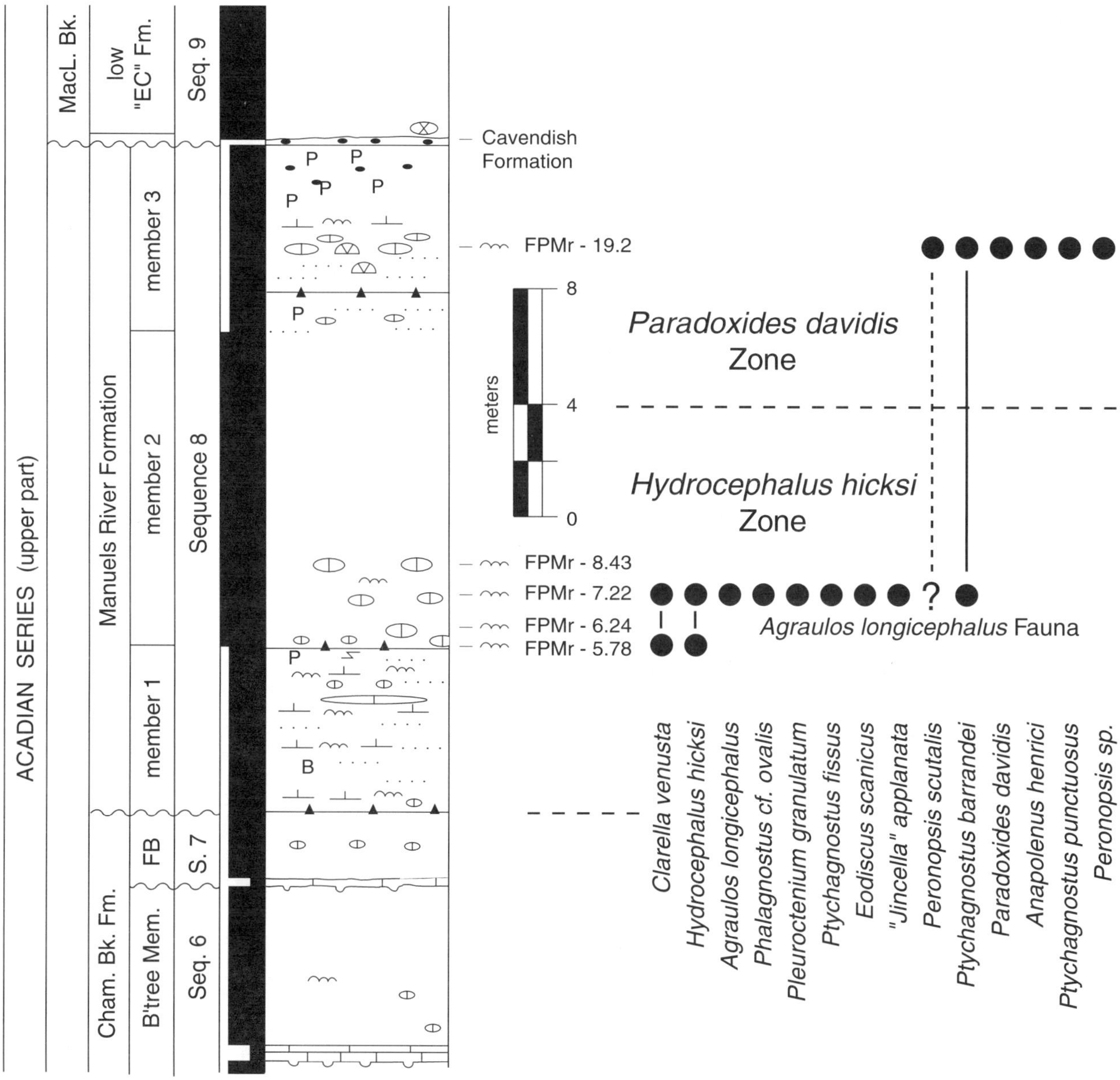

Figure 17. Acadian Series and stratigraphic ranges of trilobites and agnostoids at Fosters Point, south side of Random Island, western Trinity Bay. Abbreviations: B'tree Mem., Braintree Member; Cham. Bk. Fm., Chamberlain's Brook Formation; "EC", Elliott Cove; FB, Fossil Brook Member; MacL. Bk., MacLean Brook Group. For explanation of symbols, see Figure 6.

edge of the thicker Cavendish Formation seen at Highland Cove. The Cavendish Formation is succeeded by more highly pyritiferous, micaceous black shale with fine-grained sandstones and sideritic nodules. Poulsen and Anderson (1975) reported a lower-diversity trilobite fauna (limited to the latest Middle Cambrian agnostoids *Phalacroma falanensis* Westergaard [misidentified as *Lejopyge laevigata* (Dalman); see Rushton, 1978, and Robison, 1984] and *Peronopsis* from the lower 4 m of the MacLean Brook Group at Fosters Point) than that recovered from the same interval at

Manuels River. In addition, Hutchinson (1962, locality 20479) collected a lowest Upper Cambrian *Agnostus pisiformis* Zone fauna 18.3 m above the base of the MacLean Brook Group at this section. These data indicate a comparable development of the upper Middle Cambrian sequence 8–9 boundary at Fosters Point, Manuels River (August 12), and Highland Cove (August 13).

Sunnyside

Sunnyside: location.—The thick, gently northeast-dipping Sunnyside section forms high shore cliffs about 1 km east of Sunnyside village on the north shore of Bull Arm. The lower part of the section (Figure 18) lies below the roadside park at the top of the shore cliff and is reached by walking down the stairs to the shore. The lower part of the section can be followed eastward along the shore to the point where the section begins to trend north into West Centre Cove. The 235 m-thick Sunnyside section that has been measured extends as high as member 3 of the Cuslett Formation, and its top lies in the highland approximately 0.4 km south of the paved road (Landing and Benus, 1988b, fig. 35).

Sunnyside: upper Placentian Series.—The depositional sequence 2–3 unconformity (i.e., Random Formation–Bonavista Group contact) known along the Placentia–Bonavista axis (Figure 2) is exposed at Sunnyside. The thick sandstones at the top of the Random Formation type section at Hickman's Harbour and other localities along Smith Sound to the north (Hutchinson, 1962; Fletcher, 1972) are not present, and the thin quartz arenites and greenish silt-shales seem to lie lower in the formation. The greenish silt-shales at the top of the Random Formation become reddish and calcified 0.75 m below the Bonavista Group, and this seems to reflect weathering and initial development of a caliche (Landing et al., 1988). Subvertical quartz sand- and carbonate-filled joints that trend into horizontal fissures suggest the gentle extensional movements that led to definition of the Placentia–Bonavista axis and onlap of the Bonavista Group on the inner platform (Landing and Benus, 1988b).

A thin section (3.5 m) of the Petley

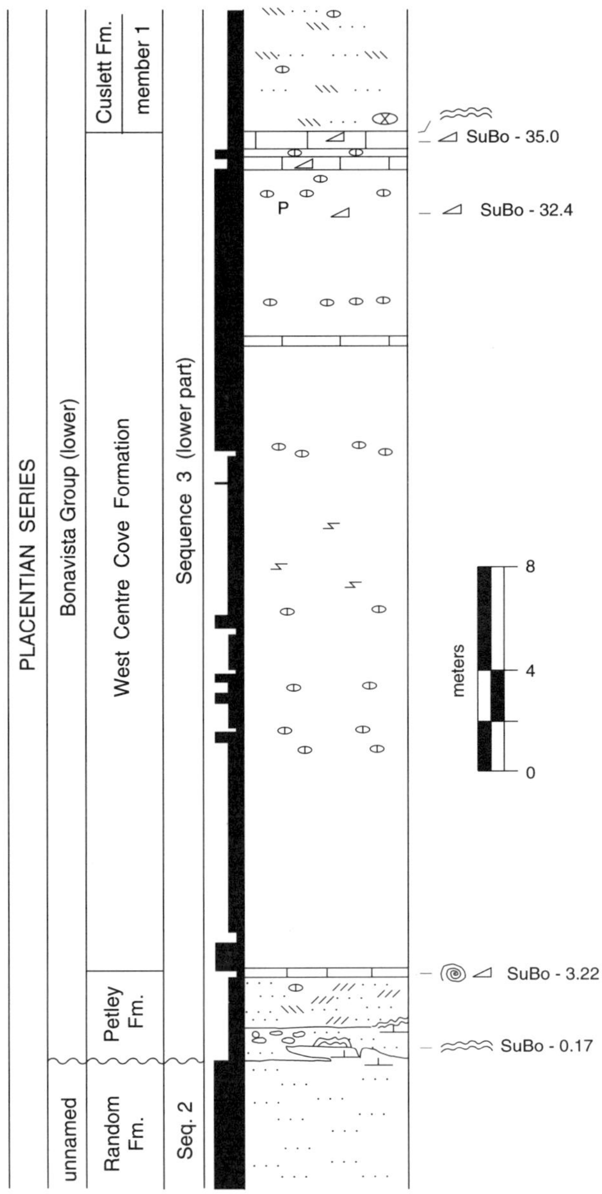

Figure 18. Lower part of the middle and upper Placentian Series at Sunnyside, Bull Arm, western Trinity Bay. For explanation of symbols, see Figure 6. Modified from Landing et al. (1988, fig. 30).

Formation, the lowest and most arenaceous formation of the Bonavista Group (Landing and Benus, 1988b), overlies the Random Formation. The Petley Formation here consists of cross-bedded, feldspathic quartz arenites; lenticular conglomerates with clasts of upper Precambrian volcanics and argillites; interbedded cyanobacterial buildups and nodular carbonates; and an uppermost 28 cm-thick fossiliferous (*Sunnaginia imbricata* Zone) wackestone and nodular lime mudstone. Five successive interchannel–tidal flat cycles (with southwest flow) are observable in the cross-bedded sandstones in the upper part of the formation. The limestones at the top of the Petley Formation have an abundant, near-shore, snail-orthothecid fauna (*Aldanella-"Allatheca"*) fauna (e.g., Landing et al., 1989) that appears with the deepening/onlap event just prior to shale deposition of the overlying West Centre Cove Formation. The genetic significance of these limestones is that they record initial West Centre Cove onlap and deepening, but they are assigned to the Petley Formation as the regional marker bed that defines the top of the Petley Formation (Landing and Benus, 1988b).

The West Centre Cove Formation type section extends as a cliff that almost reaches the mouth of West Centre Cove to the east. The top of the 32 m-thick West Centre Cove Formation is the top of the two thick orthothecid wacke- to packstone beds approximately 200 m to the east of the top of the Petley Formation. Structureless (bioturbated), red and (less abundant) green siliciclastic mudstones with diagenetic calcareous nodules and nodular limestone beds dominate the West Centre Cove Formation. Diagenetic chamosite grains and rinds (now chloritized) around the calcareous nodules are a prominent feature of the West Centre Cove Formation at this and other localities. The limestones at the top of the formation (34.0–35.4 m above the Random Formation) are separated by a 20 cm-thick interval of green siliciclastic mudstone with calcareous nodules. The middle of the upper limestone bed has limonitized, mud-cracked, planar cyanobacterial mats and thin lenses of quartz arenite with shale clasts. These features provide evidence of shoaling and onlap. The upper half of the bed has an abundant *Aldanella-"Allatheca"* fauna (i.e., *Sunnaginia imbricata* Zone) that represents reestablishment of more unrestricted marine conditions with the deepening associated with the deposition of member 1 of the Cuslett Formation. Member 1 is exposed on the west shore of West Centre Cove.

August 18: Proterozoic– Lower Acadian, Saint John, New Brunswick

The Cambrian—Lower Ordovician of the Saint John, New Brunswick region and adjacent southern New Brunswick (Figure 19) provided the basis for G.F. Matthew's classic studies of Avalonian biostratigraphy (see Miller, 1988; Landing and Miller, 1988). This cover sequence occurs in northeast-trending synclines and down-dropped fault blocks on the late Precambrian Avalonian orogen. Areas in the vicinity of Saint John as far northwest as Hanford Brook (Figure 19) have thick Precambrian–Cambrian boundary interval sections (i.e., depositional sequences 1 and 2) that are unconformably overlain by the upper Lower Cambrian Branchian Series without a record of the upper Placentian Bonavista Group (Figure 20). This stratigraphic succession demonstrates that the Saint John region is referable to the Avalonian marginal platform (Landing, 1996a).

An inner platform succession that has the feather edge of the Random Formation (depositional sequence 2) unconformable on the Avalonian orogen, and the overlying Bonavista Group (depositional sequence 3) unconformable on the Random Formation has been recognized in the Cradle Brook area to the east of Saint John (Landing, 1996b; Figure 1). A lateral transition from the western (or northern) marginal platform to eastern (or southern) inner platform comparable to that in southeastern Newfoundland and northern Wales is present in Avalonian southern New Brunswick.

The youngest age of Avalonian Cambrian–Ordovician cover sequence rocks in southern New Brunswick is late Tremadoc (see August 20, Reversing Falls). However, highly fossiliferous limestone boulders with late Arenig conodonts, trilobites, brachiopods, mollusks, and ostracodes have been found in a Triassic conglomerate in southeast Saint John (E. Landing, S.R. Westrop, and A.J. Boucot, unpub. data).

The latest Precambrian(?)–Lower Ordovician cover sequence of Avalonian New Brunswick has traditionally been referred to as the "Saint John Group" (e.g., Hayes and Howell, 1937; Alcock, 1938; McCutcheon, 1987; Currie,

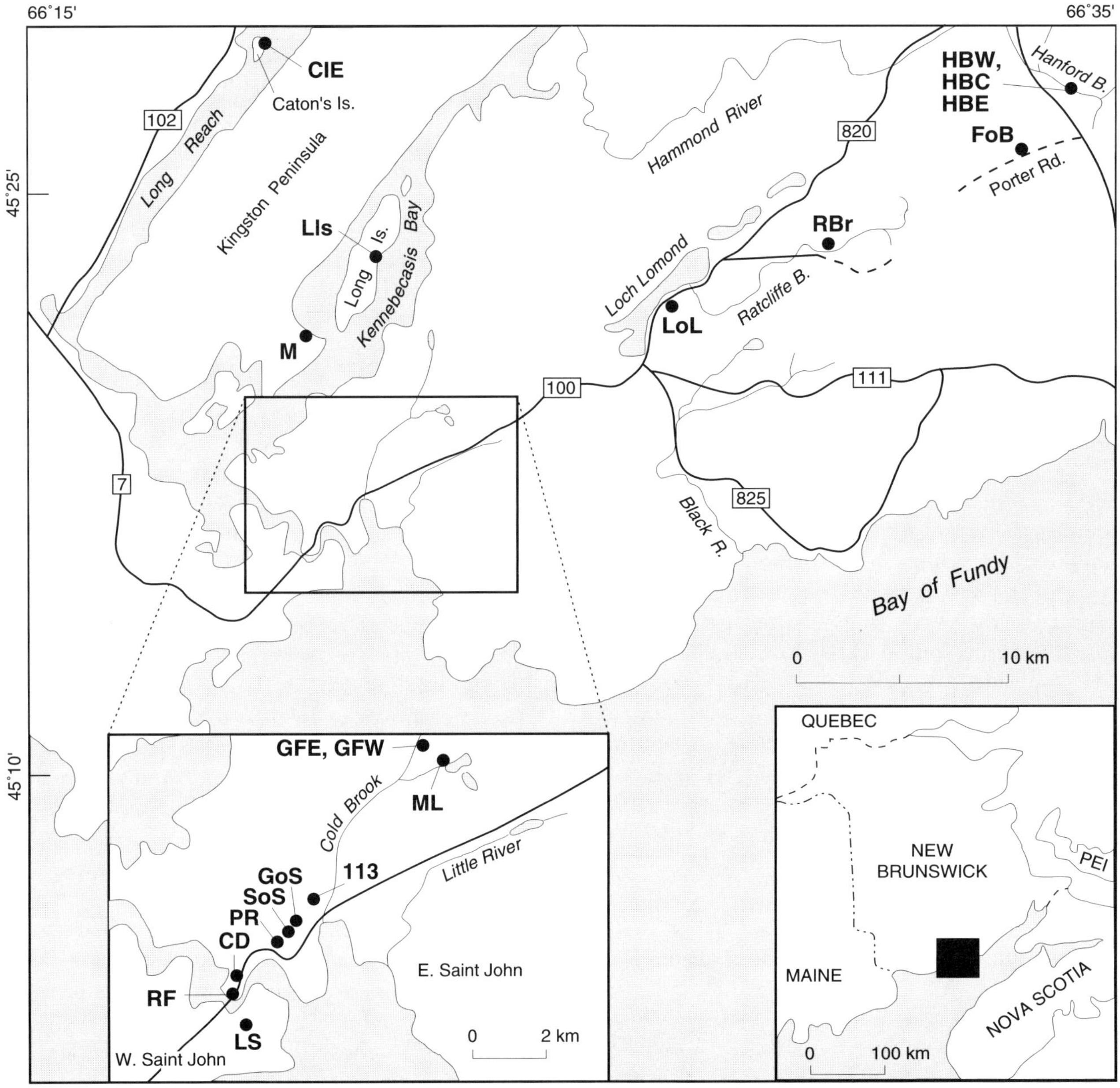

Figure 19. Uppermost Proterozoic(?)—lower Tremadoc localities in Avalonian southern New Brunswick. Abbreviations: CD, Chesley Drive; CIE, Caton's Island East; FoB, Fossil Brook; GFE and GFW, Glen Falls East and West, respectively; GoS, Gooderich Street; HBW, HBC, and HBE, Hanford Brook West, Central, and East sections, respectively; LIs, Long Island; LoL, Loch Lomond; LS, Lancaster Street, Lancaster Street; M, Milkish Head; ML, Mystery Lake; PR, Paradise Row; RBr, Ratcliffe Brook; RF, Reversing Falls; SoS, Somerset Street; 113, Exit 113 on New Brunswick Rte. 1; NS, Nova Scotia; PEI, Prince Edward Island. Broken lines are dirt roads. Modified from Landing and Westrop (1996, fig. 1).

1987; Tanoli and Pickerill, 1988). The term "Saint John Group" was abandoned by Landing (1996a) for a number of reasons: 1) it simply refers to all post-late Precambrian (i.e., Coldbrook Group volcanics) and pre-Silurian units in southern New Brunswick, and does not serve to establish a natural relationship between the units that compose it (see North American Commission on Stratigraphic Nomenclature, 1983); 2) comparison of the stratigraphy of southern New Brunswick with southeastern Newfoundland, Cape Breton Island, the northern Antigonish Highlands, and eastern Massachusetts demonstrates not only that the "Saint John Group" has a number of major unconformities that represent large intervals of geological time but is itself divisible into group-level units that are widespread in Avalon and better establish natural relationships; and 3) the original use of the designation "St. John Group" (Matthew, 1863) expressly restricted the term to the Random Formation (="Glen Falls Formation" of Hayes and Howell [1937]; abandoned by Landing, 1996a) and younger units. Hayes and Howell's (1937) expansion of the "Saint John Group" downward to include Matthew's (1863) thick "Basal group" (i.e., Rencontre and Chapel Island Formations of the Ratcliffe Brook Group; Figure 20) created an objective homonym. This is a further reason to abandon a term that does not provide a useful purpose.

Somerset Street

Somerset Street: location.—The steeply south-dipping Somerset Street section (Figure 21) extends along the east side of Somerset Street, and its top lies near the intersection of Somerset Street and Paradise Row (Figure 19, locality SoS). Somerset Street and Paradise Row are busy streets, and parking must be found on nearby streets. The Somerset Street cut has been the subject of a number of recent publications. Isachsen et al. (1993) determined a U-Pb zircon date of 530.7 ± 0.9 Ma on an ash in the upper Chapel Island Formation. Landing and Westrop (1996) dealt with the depositional history of the Hanford Brook Formation, and a U-Pb age of 511 ± 1 Ma was determined on the Hanford Brook Formation (Landing et al., In press).

Somerset Street: Proterozoic–Placentian.—Several hundred meters of massive, green-grey welded tuffs of the upper Proterozoic Coldbrook Group form the lower part of the roadcut. These tuffs are in fault-contact with the lowest sedimentary rocks of the section. The latter include 10.1 m of red arkose and feldspathic sandstones that have abundant red rhyolite pebbles. The green-red color contrast across the nearly vertical fault is obscured by hydrothermal alteration of the red color to green near the fault. These lithic sandstones are comparable to the red beds of the Rencontre Formation elsewhere in Avalon (Landing, 1996a).

The Rencontre Formation is overlain by marine siliciclastics of the Chapel Island Formation that have a distinctive basal bed (5–12 cm) of yellowish weathering, green-grey pyroclastic fragments with abundant quartz sand and quartz and shale pebbles. Tanoli and Pickerill (1988, p. 673) discussed the arguments for and against stratigraphic continuity between the red lithic sandstones and overlying marine-dominated sequence at Somerset Street and other localities, and concluded that Patel's (1977) paleosol horizon at this level is "merely composed of cleaved felsic pyroclastics". However, regional study of the Chapel Island Formation demonstrates that the lower depositional sequence of the formation (the Quaco Road Member [Landing, 1996a], known at localities to the northwest and west such as Mystery Lake, Ratcliffe Brook, and Hanford Brook [August 19]) is missing at Somerset Street. A major unconformity is present between the Rencontre Formation and Mystery Lake Member of the Chapel Island Formation (Figure 21). The basal bed of the Chapel Island Formation does indeed consist of sheared pyroclastics, but bedding-plane parallel faults occur through the section, and many weaker lithologies are sheared.

The Chapel Island Formation on the east side of Somerset Street consists of 32 m of green and minor red, purple, and dark grey shale and sandstone. Bi-directional ripples and local hummocky cross-stratification show deposition on a wave-dominated shelf, a comparable depositional environment to the Chapel Island Formation in

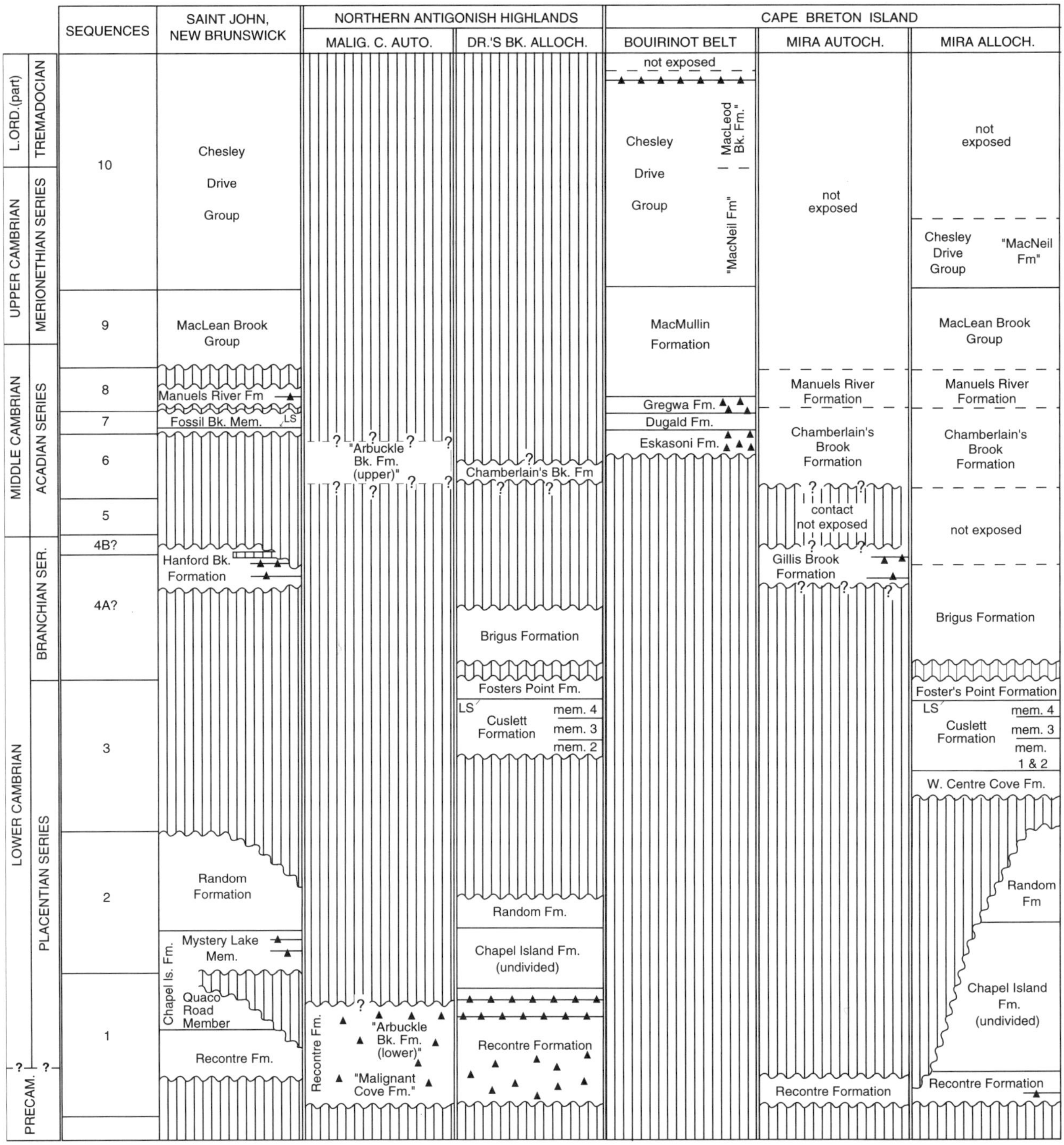

Figure 20. Uppermost Proterozoic–Lower Ordovician epeirogenic sequence stratigraphy in Maritime Canada. Modified from Landing (1996a, fig. 5).

eastern Newfoundland (Myrow et al., 1988). The contact with the overlying massive white quartz arenites of the Random Formation is faulted on the east side of Somerset Street, and an additional 8.6 m of purple and green sandstone and shale is present on the west side of the street below the Random Formation. Intercalated beds of Random-type, whitish quartz arenites in the upper Chapel Island Formation suggest a conformable transition between the two units. As in southeastern Newfoundland, the locally developed trough cross-bedding of the Random Formation shows that the transition from the Chapel Island Formation into the Random Formation corresponded to a change into a macrotidal shelf with abundant quartz sand.

Trace fossils from the Chapel Island Formation in Saint John are Early Cambrian, and include *Arenicolites* at Somerset Street (Figure 21) and *Chondrites, Palaeophycus, Skolithos,* and *Tarphrhelminthopsis* (Patel, 1985). The last four traces are present in a long road cut approximately 500 m west of Somerset Street on Paradise Row, and indicate a correlation of the Mystery Lake Member with strata of the *Rusophycus avalonensis* Zone as known in the middle–upper Chapel Island Formation in eastern Newfoundland (e.g., Narbonne and Myrow, 1988). The Mystery Lake Member at the Hanford Brook East section (August 19) has the lowest shelly fossils (the hyolith *"Allatheca" degeeri*, polychaete tubes *"Ladatheca" crosbyi, Sabellidites*) (E. Landing, unpub. data), and this indicates that the traces of the Mystery Lake Member are referable to the upper *Rusophycus avalonensis* Zone and are broadly equivalent with the Nemakit-Daldynian Stage (see Landing 1994). This interprovincial correlation means that the 530.7 ± 0.9 Ma U-Pb zircon age on an ash in the upper Chapel Island Formation at the Somerset Street cut is a pre-Tommotian date (Isachsen et al., 1993).

Somerset Street: Branchian Series.—The base of the Branchian Series and Hanford Brook Formation (emended; Landing and Westrop, 1996) consists of 1 m of dark grey quartz arenite with a phosphatic cement that is in fault-contact with the Random Formation. This "black sandstone" (Hayes and Howell, 1937) was traditionally grouped with the underlying "white quartzite" as part of the "Glen Falls Formation" (abandoned term; Landing and Westrop, 1996). Assignment of the "black sandstone" to the Hanford Brook Formation is indicated by: 1) lithologic similarity to the phosphatic sandstones to those of the St. Martins Member of the lowest Hanford Brook Formation in the Hanford Brook type area; 2)

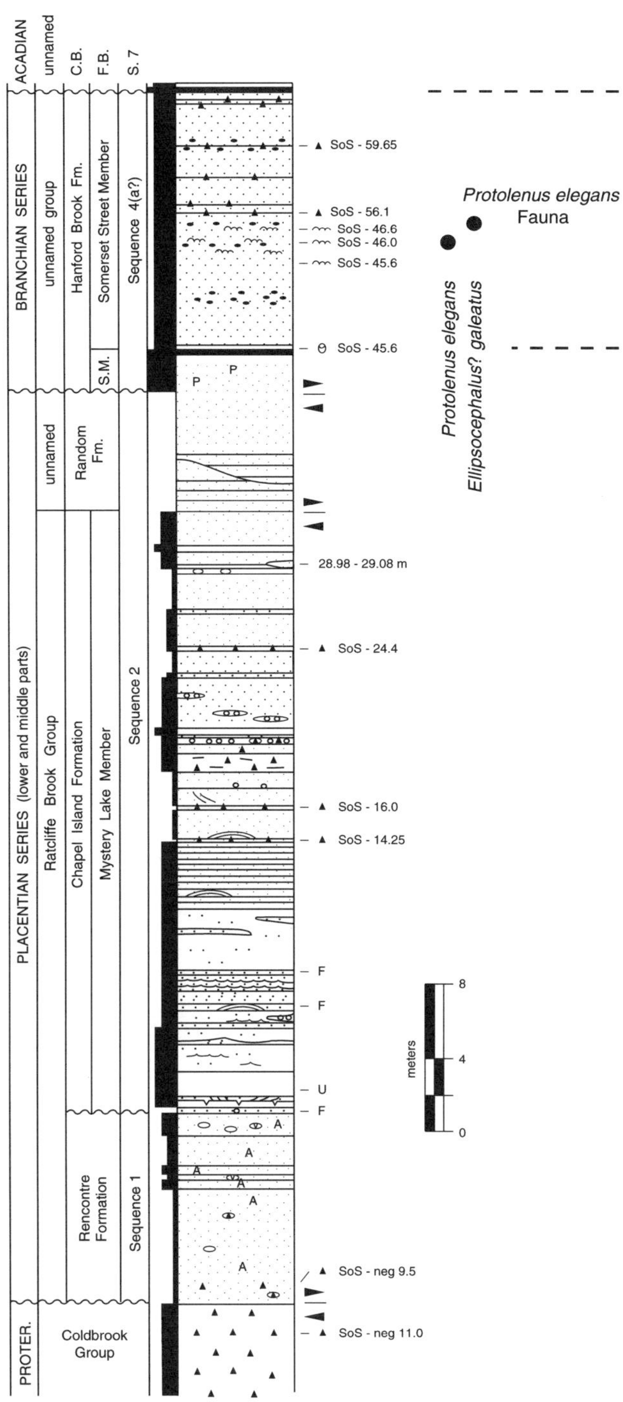

Figure 21. Proterozoic, lower–middle Placentian, and Branchian–Acadian Series at Somerset Street, Saint John, New Brunswick. Abbreviations: C.B., Chamberlain's Brook Formation; F.B., Fossil Brook Member; S.M., St. Martins Member. For explanation of symbols, see Fig. 6. Modified from Landing and Westrop (1996, fig. 3).

presence of a thick, phosphatic basal bed of the "black sandstone" on eroded "white sandstone" at a number of localities (e.g., Figure 19, locality PR); and 3) presence of the inarticulate brachiopod *"Acrotreta" gemmula* Matthew in the "black sandstone" at the Exit 113 road cut (August 20), a species characteristic of higher strata in Hanford Brook Formation and representative of a group of inarticulates unknown in the Placentian Series (Landing and Westrop, 1996). These lithostratigraphic and biostratigraphic data mean that the Random—Hanford Brook interformational contact is an unconformity between the lower Placentian Series (depositional sequence 2) and the Branchian Series (depositional sequence 4). This unconformity is comparable to the Random Formation–Brigus Formation unconformity in the southern Burin Peninsula, southeastern Newfoundland, a region which also lies on the Avalonian marginal platform (August 14, Little Dantzig Cove Brook).

Higher strata of the Hanford Brook Formation at the Somerset Street roadcut comprise the Somerset Street Member (type section), and include meter-scale alternations of medium grey and more pyritiferous, dark grey, fine-grained sandstones with prominent planar laminations and small burrows. These features are consistent with deposition in a low energy, dysaerobic environment (Landing and Westrop, 1996). The highest unit of the Hanford Brook Formation, the Long Island Member (see August 19, Hanford Brook West), has been eroded below the unconformity with the overlying Chamberlain's Brook Formation. The Somerset Street section has yielded trilobites. *Protolenus elegans* Matthew and *Ellipsocephalus*? *galeatus* Matthew occur in the middle of the Somerset Street Member (Figure 21). This assemblage characterizes the *Protolenus elegans* Fauna in the Somerset Street Member at Hanford Brook (section HBW, August 19). Volcanic ashes occur through the middle and upper Somerset Street Member, and the ash located at 56.1 m above the Coldbrook Group (Figure 21) has yielded a 511 ± 1 Ma age on strata correlated into the middle Botomian–Toyonian stages of Siberia (Landing et al., In press).

Somerset Street: Acadian Series.—A 25 cm-thick, arenaceous phosphate bed with phosphate pebbles to 3x2 cm in diameter and 10 cm of overlying dark phosphatic shale comprises the lowest beds of the Fossil Brook Member of the Chamberlain's Brook Formation. Fossil Brook Member faunas from the Gooderich Street cut just east of Somerset Street (Figure 19, locality GS; S.R. Westrop and E. Landing, unpub. data), from the Exit 113 section (August 20), Fossil Brook (Figure 19, locality FOB, see "Biostratigraphic and geochronologic overview," above), and from Hanford Brook West (August 19) are all referable to the *Paradoxides eteminicus* Zone and depositional sequence 7. This means that the Hanford Brook—Chamberlain's Brook contact is a major unconformity between depositional sequence 4 and 7. This unconformity represents a hiatus significantly longer than that locally seen between the Lower and Middle Cambrian on the marginal platform in the southern Burin Peninsula, southeastern Newfoundland (August 14, Little Dantzig Cove Brook).

AUGUST 19: PROTEROZOIC–ACADIAN OF HANFORD BROOK

Hanford Brook

Hanford Brook localities.—The well-exposed Proterozoic–Acadian sections on Hanford Brook (Figures 19, 22–24) lie in the east limb of a broad Cambrian syncline approximately 14 km north of the village of St. Martins. The localities (Figure 19, HBE, HBC, and HBW) are reached by crossing private land south of the hamlet of Hanford Brook. Vehicles will exit Rte. 111 onto a long gravel driveway marked with the family name "Allaby," and will be parked near the garage. Hanford Brook is reached by walking directly north for several hundred meters along the edge of the woods.

The Hanford Brook sequence dips downstream, and field trip participants will walk several kilometers upstream to the base of the Cambrian succession (Figure 22). Examination of the Hanford Brook sections will begin at the base of the succession. The Branchian–Acadian sequence of the Hanford Brook West (HBW) succession (Figure 24) is traversed first in the walk to the Precambrian–Cambrian boundary (Coldbrook Group–Rencontre Formation) unconformity.

Middle Placentian quartz arenites of the Random Formation with a cap of Branchian Series conglomeratic sandstone are separated by a fault from the HBW section and form a high, steeply dipping (70° N) cliff on the northeast side of Hanford Brook. This Random Formation-dominated cliff just upstream from HBW is the Hanford Brook Central (HBC) section (Figure 23).

A second fault that crosses Hanford Brook separates the Random Formation in section HBC from gently northwest-dipping (30°) quartz arenite ledges of the Random Formation that appear only 15 m upstream from the HBC section. These gently dipping Random ledges form the top of the thick (378 m) Hanford Brook East (HBE) succession (Figure 22). A 286.5 m-thick section from the base of the Rencontre Formation through the exposed Chapel Island Formation (286.5 m) is illustrated herein (Figure 22). A higher covered interval (286.5-335 m) and a section in the Random Formation (335-378 m) are not shown in Figure 22.

Hanford Brook East (HBE): Rencontre–Chapel Island formations.—A 34.75 m-thick section of arkose fanglomerates that are transitional upward into subaerial, feldspathic, quartz pebble conglomerates unconformably overlies purple rhyolites of the late Proterozoic Coldbrook Group. A fault oblique to bedding cuts the upper part of this conglomeratic sequence, and the top 0.35–2.15 m of the conglomerate is locally bleached white.

Thin- to medium-bedded, trough cross-bedded feldspathic arenites overlie the conglomerates and represent an abrupt transition from red beds comparable to the Rencontre Formation of southeastern Newfoundland into a marine facies comparable with the lower Chapel Island Formation. Phosphatic pebbles, wave ripples, and calcareous nodules appear, respectively, at 42.75 m, 44.7 m, and 48.4 m above the Coldbrook Group volcanics and show the establishment of shallow marine conditions. Large looping trace fossils (i.e., *Tarphrhelminthoida* Ksiazkiewisz) in wave-rippled sandstones at 51 m (Hofmann and Patel, 1988) show the behaviorly complex activity of Early Cambrian metazoans. Absence of fossils below these traces makes correlation of the underlying Rencontre Formation impossible. However, the relatively minor stratigraphic thickness of these rapidly deposited fanglomerates below the Cambrian traces suggests that their deposition was not initiated in the late Proterozoic. It might be noted here that calcareous sandstones and calcareous nodules (samples HBE-48.8, -77.85, -81.35) did not yield small shelly fossils by crack-out or acid disaggregation, and correlation of this part of the section into the lowest Placentian Series below the lowest small shelly fossils in southeastern Newfoundland seems likely. Volcanic ashes that we have collected for U-Pb zircon dates through the lower 92 m of the Chapel Island Formation at Hanford Brook East may help refine correlations.

An important stratigraphic break is indicated by conglomerate-filled channels in the interval 157.5–174.3 m. These stream-laid gravels overlie the wave-dominated (frequently hummocky cross-stratified) sandstones of the lower Chapel Island Formation. These channel fills indicate incision associated with development of a type 1 sequence boundary. Unburrowed, olive green shales with centimeter-thick glauconitic sandstone lenses and large methanogenic nodules abruptly overlie the conglomerates, and record a subsequent strong marine onlap. Landing (1996a) used the base of this conglomeratic interval to define: 1) the depositional sequence 1–2 unconformity within the lower Placentian Series in southern New Brunswick, and 2) the contact of the two Chapel Island Formation members–a Quaco Road Member and an upper Mystery Lake Member.

The lowest shelly fossils in Avalonian New Brunswick occur in calcareous nodules that are developed in a shallower facies (red-brown, fine-grained sandstones) above the base of the Mystery Lake Member (samples HBE-191, -193.75). Matthew (1889) reported body fossils from this interval. *"Ladatheca"* conchs, the orthothecid *"Allatheca"*, and *Sabellidites* from this part of the section (E. Landing, unpub. data) sup-

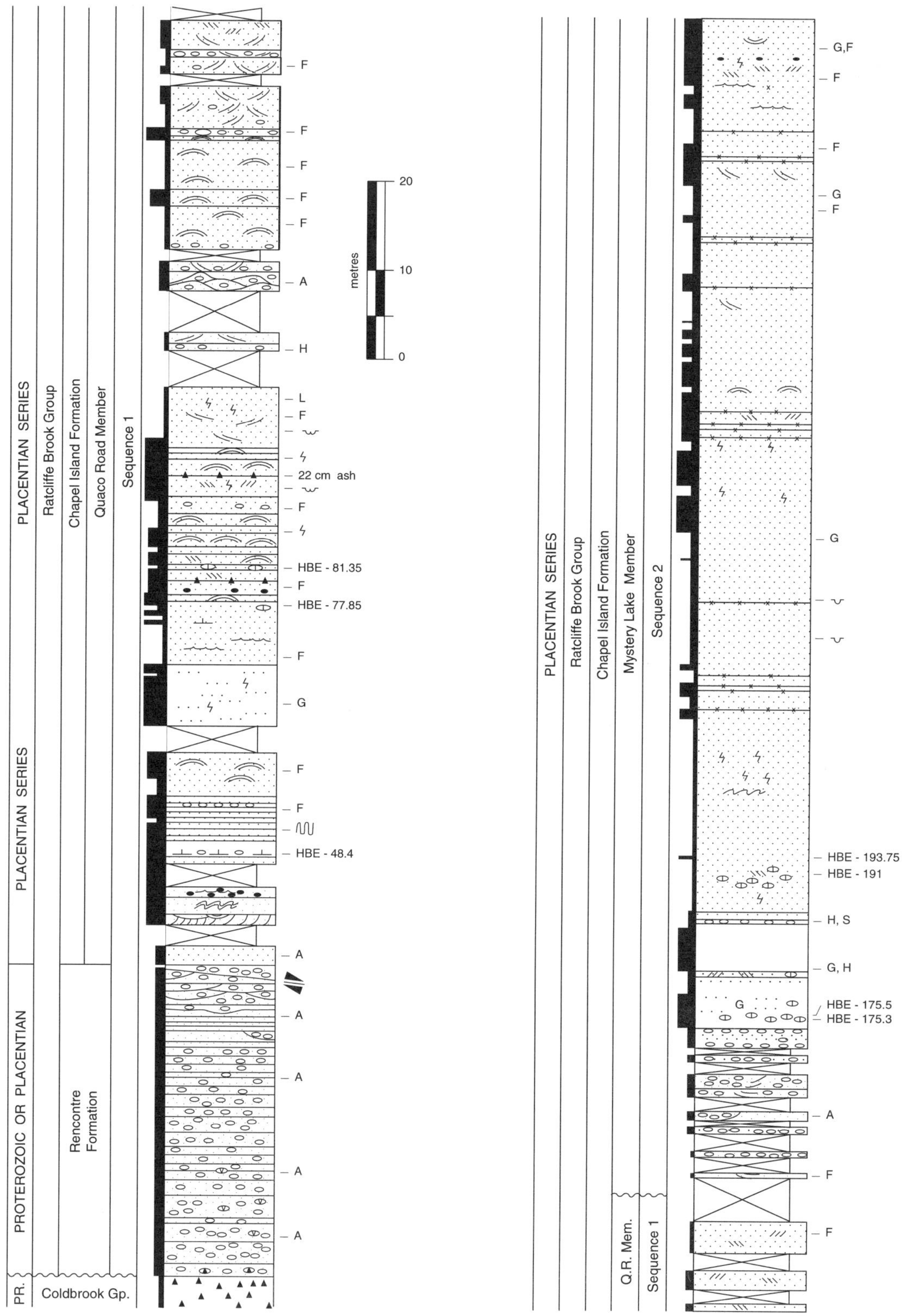

Figure 22. Proterozoic and lower Placentian Series to top of exposed Chapel Island Formation in Hanford Brook East (HBE) section. As noted in text, an overlying 48.5 m covered interval (286.5–335 m) and the exposed Random Formation (335–378 m) are not shown. For explanation of symbols, see Figure 6.

port a correlation with the *Watsonella crosbyi* Zone.

Biostratigraphically useful fossils have not been recovered from the upper Chapel Island Formation at Hanford Brook East. These thin- to medium-bedded, red to green, feldspathic sandstones are locally churned by *Teichichnus* Seilacher and characterized by wave-produced sedimentary structures. Thin, highly siliceous (cherty), fine-grained sandstones in this part of the section may be altered volcanic ashes.

Hanford Brook Central (HBC): Random–Hanford Brook unconformity.—The Hanford Brook Central section (Figure 23) is dominated by 31.6 m of steeply dipping; medium- to coarse-grained; pink, buff, or light green; somewhat feldspathic quartz arenites with planar to trough cross-beds. Deposition was under E–W tidal currents. The base of this macrotidal sandstone facies is interbedded with the uppermost, wave-deposited sandstones of the Chapel Island Formation on marginal platform sites elsewhere in southern New Brunswick (August 18, Somerset Street), southeastern Newfoundland (August 14, Little Dantzig Cove), Cape Breton Island (Landing, 1991), and the northern Antigonish Highlands (Landing and Murphy, 1991). All of these macrotidal sandstone occurrences are referred to the middle Placentian Random Formation (Landing, 1996a).

Medium- to massive-bedded, rusty-weathering, hematitic, glauconitic quartz arenites with trough cross-beds (31.6–33.75 m) overlie the Random Formation in the woods at the west end of Hanford Brook Central. These hematitic sandstones have large rounded clasts of coarse-grained quartz arenite (to 10x15x8 cm) and tabular clasts of finer-grained sandstone (Landing and Westrop, 1996, fig. 5g). This sandstone-on-sandstone contact is the unconformable contact of the St. Martins Member of the Hanford Brook Formation (depositional sequence 4) on the Random Formation (depositional sequence 2).

Hanford Brook West (HBW): Branchian Series, Hanford Brook Formation, St. Martins Member.—The type section of the Hanford Brook Formation (Hayes and Howell, 1937) is 32.75 m thick (Figure 24). Landing and Westrop (1996) divided the type section into three successive members and

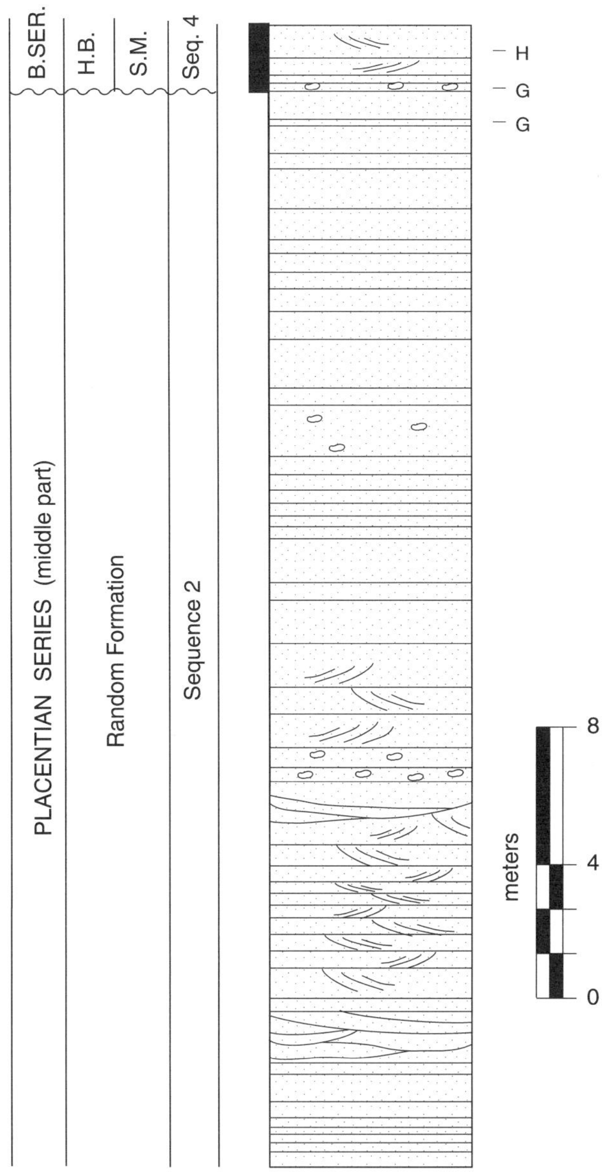

Figure 23. Middle Placentian and lower Branchian at Hanford Brook Central (HBC). For explanation of symbols, see Figure 6.

described depositional environments.

The lower 12 m at HBW is the type section of the St. Martins Member of the lower Hanford Brook Formation. The contact of the St. Martins Member with the Random Formation is exposed at the top of the Hanford Brook Central (HBC) section (described above), but is covered, and likely faulted, below the HBW succession. The thickest known section of the St. Martins Member is at HBW, where it consists of 20—90 cm-thick intercalations of blue-grey, fine-grained sandstones with prominent burrowing and dark green-grey to black, medium- to very coarse-grained, glauconitic (now chloritic) quartz arenites with prominent trough cross-beds and phosphate-impregnated caps.

These near-shore alternations of low- and episodically high-energy sandstones are Hayes and Howell's (1937) "*Beyrichona* beds", with a fauna dominated by secondarily phosphatized "ostracodes" (*Beyrichona* and *Hipponicharion*) and lingulelloid and acrotretid brachiopods. Although Matthew (1895) noted two trilobites of the *Protolenus* fauna from this lower part of the Hanford Brook Formation, subsequent workers (Hayes and Howell, 1937; Alcock, 1938; Landing and Westrop, 1996) did not find trilobites. This apparent lack of trilobites is interpreted to be either a result of dissolution of calcareous remains in the non-calcareous muds and sands of the St. Martin's Member, or an absence of trilobites in very near-shore Lower Cambrian deposits. By the latter interpretation, the St. Martins Member is an ostracode-brachiopod biofacies of the ordinarily trilobite-bearing Branchian Series (Landing, 1992). This ostracode-brachiopod biofacies is also seen in the Ab_2 and Ab_3 units of the Lower Comley Sandstone in Shropshire (*Acrothele prima* Shale and green sandstone, respectively; see Rushton, 1974) and in the Dugald Formation of Cape Breton Island (Hutchinson, 1952), which also lack or have very rare trilobites.

Hanford Brook West (HBW): Branchian Series, Hanford Brook Formation, Somerset Street–Long Island Members (revised).—The overlying Somerset Street Member of the Hanford Brook Formation is dominated by dark grey, fine-grained sandstones that are only weakly burrowed and generally laminated. Trilobites of the *Protolenus elegans* Fauna occur with the ostracodes described above from the St. Martins Member. The *Protolenus*-bearing fauna of the Hanford Brook Formation serves as the "type" assemblage for a so-called "*Protolenus* Zone" that is generally regarded as uppermost Lower Cambrian in Avalonian and Baltic successions (Cowie et al., 1972).

This lower-energy, dysaerobic facies represents the high-stand systems tract of the lower Hanford Brook Formation. Its base at 12 m in the measured section is a condensed phosphatic pebble lag deposit that includes the well-preserved, strongly convex phosphatic steinkerns that Matthew (1895) used to diagnose his genus *Protolenus* (Landing and Westrop, 1996).

The base of the upper member, or Long Island Member of the Hanford Brook Formation (Landing and Westrop, 1996), is revised at HBW by lowering it to the base of the massive (20.25–21.45 m), glauconitic (now metamorphosed to chlorite), coarse-grained sandstone bed. This revision of the Somerset Street and Long Island Members places their contact at a stratigraphic break between blue-grey, fine-grained sandstones and an overlying phosphate pebble-bearing sandstone that has isolated pebbles and granules of Proterozoic vein quartz and rhyolite (Landing and Westrop, 1996). Elements of the *Protolenus elegans* Fauna occur in phosphatic pebbles at the base of the sandstone and as isolated specimens at the top of the sandstone. This biostratigraphic evidence indicates that the stratigraphically abrupt appearance of this higher-energy sandstone and erosion of the Somerset Member to yield fossiliferous phosphate clasts was within the *Protolenus elegans* Fauna interval, and may not represent a long hiatus.

Our interpretation of the revised Somerset Street–Long Island Member contact is as a Branchian Series offlap–onlap event comparable to and possibly coeval with the boundary of the St. Mary's and Jigging Cove Members of the Brigus Formation (i.e., sequences 4A and 4B) in eastern Newfoundland (e.g., August 14, Jigging Cove, and August 16, Smith Point). Because depositional sequence boundaries are essentially synchronous across Avalon (Landing, 1996a), and because the sequence 4A–AB boundary corresponds to the contact of the *Callavia broeggeri* Zone and overlying "*Myopsostrenua sabulosa* Zone" in southeastern Newfoundland, a temporal correlation of the Somerset Street–Long Island contact with the St. Mary's–Jigging Cove contact would require a reevaluation of the biostratigraphic significance of *Protolenus*-bearing assemblages. Such a correlation would mean that the classic

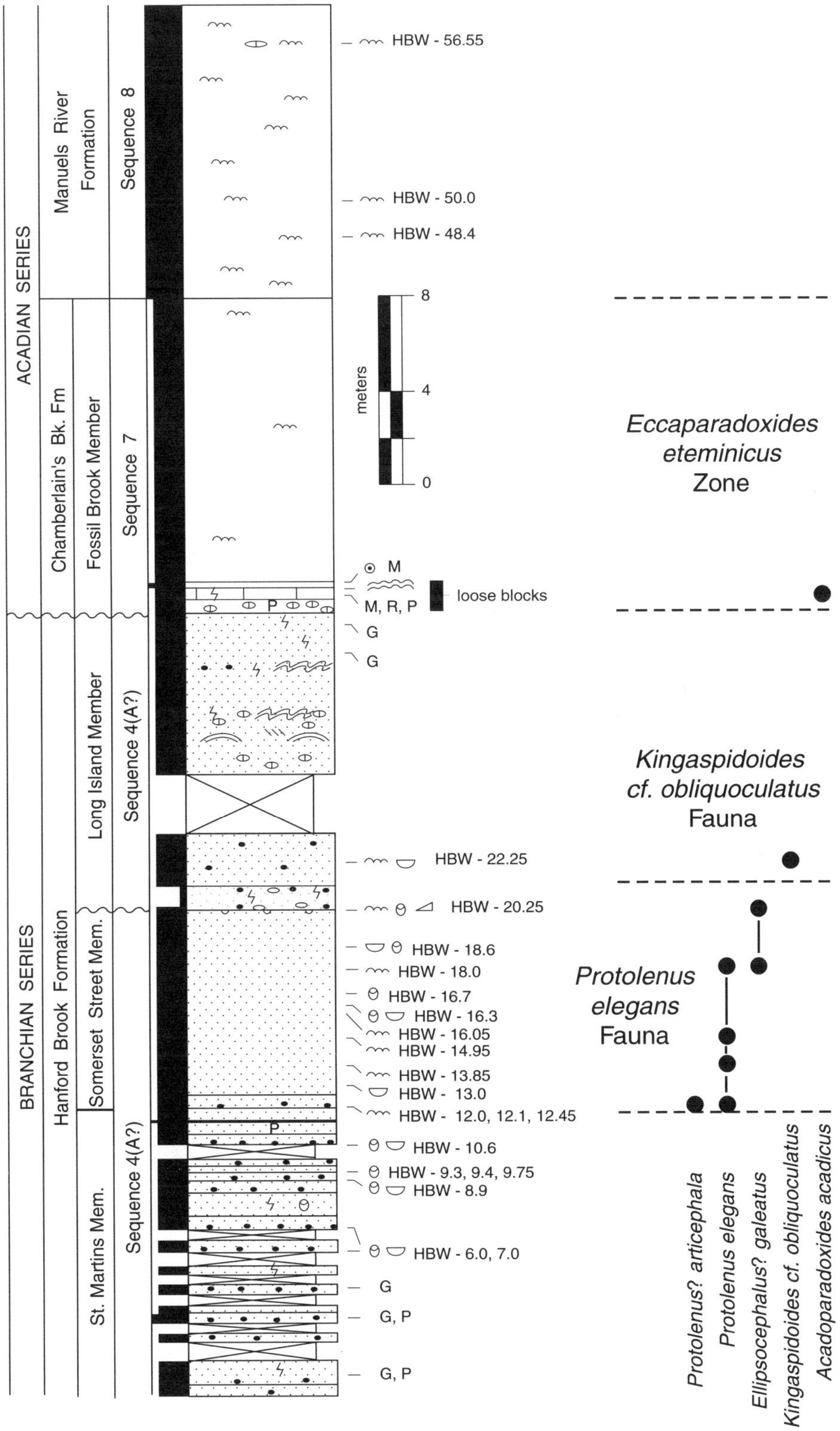

Figure 24. Branchian–Acadian Series and ranges of trilobites at Hanford Brook West (HBW) section. For explanation of symbols, see Figure 6. Modified from Landing and Westrop (1996, fig. 2).

"*Protolenus* Zone" of the Hanford Brook Formation would not be high in the trilobite-bearing Lower Cambrian (e.g., Geyer and Palmer, 1995), but would be a faunal assemblage in sandstone facies that would correlate with the *Callavia broeggeri* Zone in siliciclastic mudstones and nodular limestones.

The massive sandstone of the lowest Long Island Member is overlain by 2 m of dark grey, fine-grained sandstone similar to that of the underlying dysaerobic, low-energy facies of the Somerset Street Member. These latter beds are followed by a 2.5 m covered interval, and then by a cliff of fine-grained, thin-bedded, glauconitic (now chlorite), planar-bedded to hummocky cross-stratified, more strongly burrowed sandstone. This episodically high-energy, more oxygenated, wave-dominated facies represents the highstand systems tract of the second Hanford Brook subsequence.

Hanford Brook West (HBW): Hanford Brook Formation trilobites.—The trilobites of the Hanford Brook Formation have been assigned to a "*Protolenus* Zone" (Hayes and Howell, 1937), but our collections indicate that this zone is equivalent to only the lower assemblage of the formation (Figure 24). This interval is referred to herein as the *Protolenus elegans* Fauna, and is characterized by the eponymous species and rare representatives of *Protolenus*? *articephalus* (Matthew; = *Avalonia acadica* Matthew and *Protagraulos priscus* Matthew) and *Ellipsocephalus*? *galeatus* Matthew. The unique specimens of *Micmacca matthevi* Matthew (?=*M. vaningeni* Matthew) and *Mohicana plana* (Matthew) were also collected from this stratigraphic interval (Matthew, 1895) but were not encountered during our field work.

The younger assemblage is assigned to a newly designated *Kingaspidoides* sp. cf. *K. obliquoculatus* Fauna of the Long Island Member. This fauna is poorly represented at Hanford Brook, but occurs at Long Island (section LIs; Figure 19). Here, the eponymous species occurs with *Berabichia* n. sp., the only other form known in the fauna.

The entire "*Protolenus* Zone" fauna consists of species endemic to the Hanford Brook Formation, and accurate biostratigraphic correlation is difficult. Correlations based on comparisons at the generic level equate the "*Protolenus* Zone" with the Tissafinian Stage of Morocco (Landing et al., in press), and conclude that it is younger than the *Callavia broeggeri* Zone in Newfoundland. However, as noted above, sequence stratigraphic criteria raise the alternative hypothesis that the faunas of the Somerset Street Member are actually the sandstone biofacies-equivalent of *Callavia broeggeri* Zone faunas. At present, there are insufficient data to choose between these alternatives.

Hanford Brook West (HBW): Acadian Series, Chamberlain's Brook Formation.—The light grey-green, siliciclastic mudstone-dominated Fossil Brook Member (depositional sequence 7) of the Chamberlain's Brook Formation overlies the Hanford Brook Formation (depositional sequence 4) across southern New Brunswick (Landing, 1996a). Most of the Fossil Brook Member (32.75–46 m) is exposed only at very low water in Hanford Brook, and its paradoxidiid-dominated fauna is difficult to collect. However, the lithologically complex lowest beds form a resistant ledge above the Hanford Brook Formation.

The Hanford Brook Formation is overlain by a resistant green siltstone (32.75–33.2 m) with abundant calcareous nodules, scattered phosphatic nodules, and cubic pyrite. Directly above this siltstone is an even more resistant ledge (33.2–33.85) of chlorite-, phosphate-, and manganese oxide-replaced, nodular trilobite wacke- to packstone with phosphatic and rhodochrosite nodules and an upper 0–20 cm-thick lens of manganese oxide-impregnated SH-V stromatolites that sit on a caliche horizon. Uncompacted, well-preserved specimens of *Acadoparadoxides acadicus* (Matthew) occur in the middle of this upper ledge in an isolated outcrop on the north side of Hanford Brook (S.R. Westrop and E. Landing, unpublished data). A 15 cm-thick bed of very dense, black, manganiferous siliciclastic mudstone overlies the stromatolites.

As noted by Howell (1925) and Fletcher (1972), Fossil Brook Member trilobites are comparable to that of the uppermost Chamberlain's Brook Formation in southeast Newfoundland (e.g., August 12, Red Bridge Road). This

Eccaparadoxides eteminicus assemblage (see discussion of Porter Road locality) appears in nodular limestones immediately above Hanford Brook Formation–Fossil Brook Member contacts in Saint John, New Brunswick and the Hanford Brook outlier, and emphasizes the considerable length of the upper Lower Cambrian–middle Middle Cambrian (depositional sequence 4–7) unconformity in Avalonian New Brunswick (Figure 20). The manganiferous shale, rhodochrosite nodules, and manganiferous carbonates in the lowest Fossil Brook Member at Hanford Brook and elsewhere in southern New Brunswick (E. Landing, unpub. data) are features common to the Chamberlain's Brook Formation in eastern Newfoundland and the northern Antigonish Highlands, northern Nova Scotia (Hutchinson, 1962; Landing, 1995).

Hanford Brook West (HBW): Manuels River Formation.—A 12 m-thick section (46–58 m) of dark grey siliciclastic mudstone overlies the Fossil Brook Member, but can only be observed in the bed of Hanford Brook at very low water. Trilobite remains are abundant throughout the unit and have been collected from thin tempestite horizons and a calcareous nodule. An abrupt change from light grey of the Fossil Brook Member to a much darker grey color permits recognition of the Fossil Brook Member– Manuels River Formation contact. The top of the Manuels River Formation is not exposed in Hanford Brook.

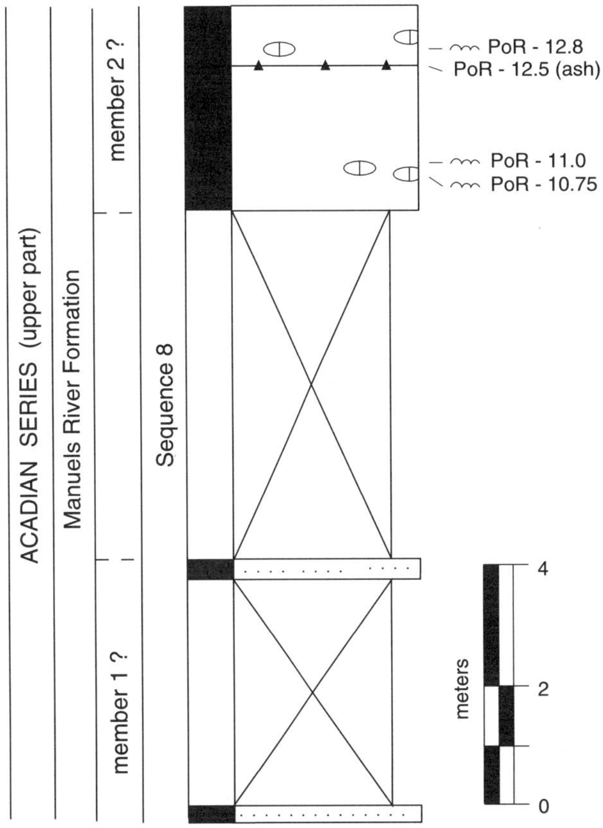

Figure 25. Upper Acadian Series at Porter Road. For explanation of symbols, see Figure 6.

Porter Road

Porter Road: location.—Acadian Series black shale lies in the bed of Porter Road. The locality is reached by turning west from Rte. 111 onto Porter Road (unpaved and unmarked) and driving for approximately 4 km. Porter Road lies in the east limb of the same Cambrian syncline as the Hanford Brook sections. Scattered outcrops along the dirt road include ledges of Rencontre Formation red beds near the intersection with Rte. 111; low outcrops of Chapel Island Formation sandstones occur further west. Vehicles will be parked at the top of the hill that descends into the broad Porter Brook lowland. The brook that passes under Porter Road in a culvert at the top of the hill is Fossil Brook (Hayes and Howell, 1937). The Fossil Brook canyon has a section through the Hanford Brook Formation into the type section of the Fossil Brook Formation of Hayes and Howell (1937; revised to Fossil Brook Member of the Chamberlain's Brook Formation by Landing, 1996a). The Fossil Brook Member has a diverse *Eccaparadoxides eteminicus* Zone assemblage with *E. eteminicus* (Matthew), *Acadoparadoxides acadicus* (Matthew), *Braintreella* sp., *Hartella matthewi* (Hartt), *Bailiaspis baileyi* (Hartt), and the eodiscoid *Dawsonia dawsoni* (Hartt).

Porter Road: Acadian Series, Manuels River Formation.—Low outcrops in the bed and on the south edge of Porter Road were the type locality and (at that time) only fossiliferous outcrops of Hayes and Howell's (1937) Porter Road Formation. The top and bottom contacts of the formation are not exposed here. The type locality of the older Fossil Brook Member (Fossil Brook

"Formation" of Hayes and Howell, 1937; revised to "Member" by Landing, 1996a) of the upper Chamberlain's Brook Formation, or depositional sequence 7 of the Avalonian Cambrian, is about 200 m northeast. Hayes and Howell (1937) reported arenaceous dark shales of their "*Goniagnostus confluens* shale" in higher outcrops near the foot of the hill.

The lithology at the Porter Road Formation type section and other localities of the formation consists of dark grey to black shale with locally large calcareous nodules and thin volcanic ash (Figure 25). Abundant trilobites, which are dominated by agnostoids in earlier reports and include the topotypes of *Paradoxides abenacus* Matthew (C.F. Hartt *in* Dawson, 1868, p. 655–656; Matthew, 1885, 1896), occur in the section. Calcareous nodules at this section have yielded several hundred uncompacted sclerites, which include *P. abenacus* (=*P. walcotti* Howell) and *"Jincella" acadica* (Whiteaves) (S.R. Westrop and E. Landing, unpub. data). Numerous cephala and pygidia of the agnostoid *Hypagnostus parvifrons* (Linnarsson) occur in all of the nodules, along with a few specimens of *Ptychagnostus perrugatus* (Gronwall) and *Goniagnostus* sp.

Curiously, the eodiscoid *Eodiscus scanicus* (Linnarsson), numerous sclerites of which are associated with the type material of *P. abenacus* (preserved in shale), does not occur in the nodules. Similarly, the types of *Peronopsis acadicus* (Hartt; see Robison, 1995) are also preserved in shale, but this species is absent from the nodules. The occurrence of *E. scanicus* in the shales at Porter Road indicates a correlation with the *Hydrocephalus hicksi* Zone in southeast Newfoundland (see also Hayes and Howell, 1937), a conclusion which is also supported by the presence of *P. perrugatus* in the nodules. However, the Porter Road fauna is dominated by species that are either rare (*H. parvifrons*) or unrecorded (*P. abenacus*; *"J." acadica*) from Newfoundland, and this suggests the existence of biofacies differentiation within the widespread shale basin of the Manuels River Formation.

After severe rains, about 4 m of black siliciclastic mudstone are exposed on the north shoulder of Porter Road and approximately 12 m above the section in Figure 25. Moldic, poorly preserved ptychagnostids and fragments of a paradoxidiid trilobite were recovered from the base of this low exposure (E. Landing and S.R. Westrop, unpublished data). This mudstone is lithologically and faunally referable to the Manuels River Formation.

All data on stratigraphic position, lithology, and faunas of the "Porter Road Formation" (abandoned by Landing, 1996a) allow confident assignment of this "formation" to the Manuels River Formation (Landing, 1996a). Tanoli and Pickerill's (1988) proposal of a "Forest Hills Formation" for an interval corresponding to Hayes and Howell's (1937) "Fossil Brook Member" and "Porter Road Member" is unwarranted. These two units, a lower green shale-dominated unit with basal manganiferous carbonate or trilobite packstones and an upper black shale-dominated unit with methanogenic carbonates and volcanic ashes, are readily distinguishable in outcrop (e.g., Hayes and Howell, 1937; Alcock, 1938; Landing and Westrop, 1996).

AUGUST 20: PROTEROZOIC–LOWER ORDOVICIAN, SAINT JOHN, NEW BRUNSWICK

Reversing Falls

Reversing Falls: location.—A steeply south-dipping, black siliciclastic mudstone-dominated section with sideritic nodules (Figure 26) is exposed on the north side of the Saint John River immediately downstream from the Rte. 100 bridge. Parking is available across the river at the Reversing Falls overlook park.

Reversing Falls: upper Tremadoc sequence and faunas.—Matthew (1892) first described the olenid trilobite *Cyclognathus rotundifrons* (now referred to *Peltocare* Henningsmoen; see Henningsmoen, 1957) from the Reversing Falls section, and McLearn (1915) described graptolites from the section. McLearn's graptolite identifications and his correlation of the Reversing Falls section into the Arenig have been followed by subsequent workers (e.g., Hayes and Howell, 1937; Tanoli and Pickerill, 1988). This traditional assignment has been re-evaluated.

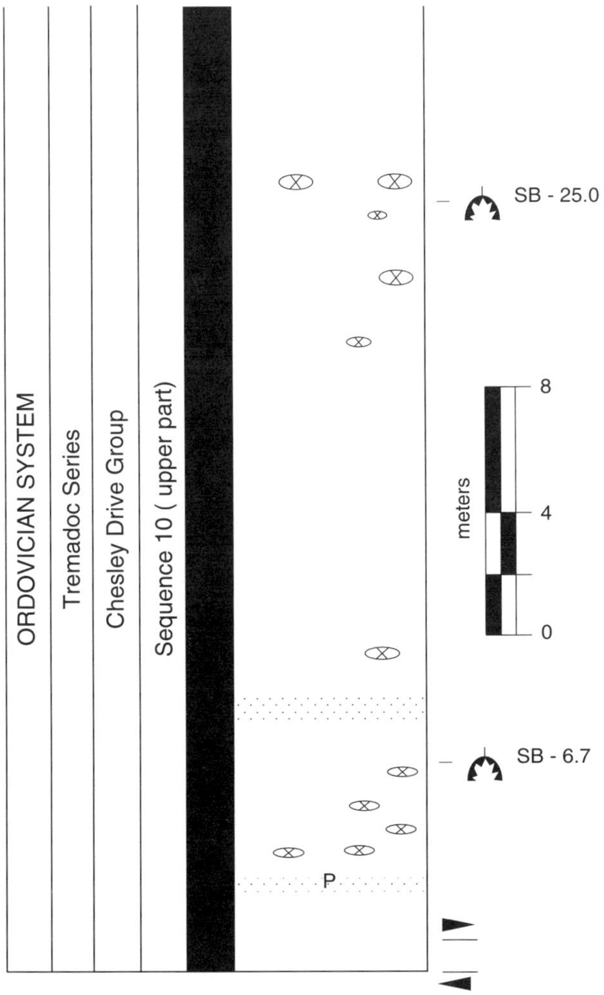

Figure 26. Upper Tremadoc Series at Reversing Falls, Saint John, New Brunswick. For explanation of symbols, see Figure 6. Modified from Landing (1980, fig. 3).

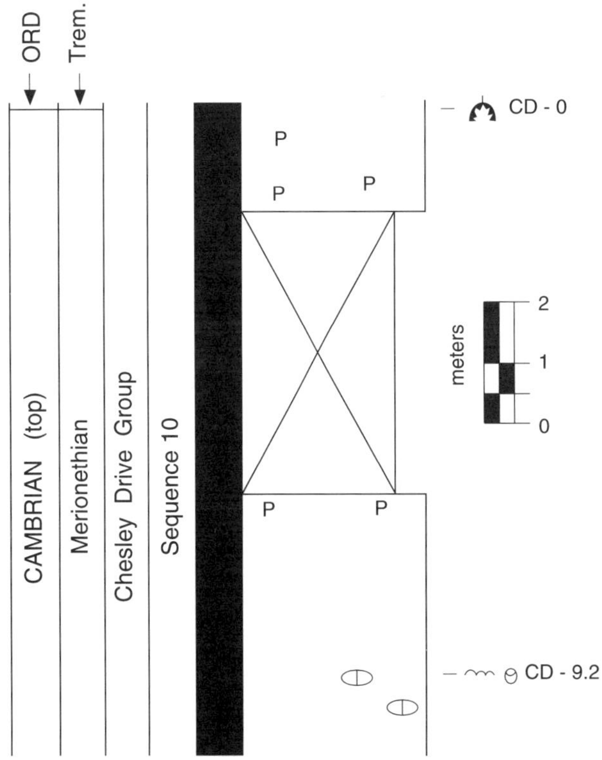

Figure 27. Merionethian–lowest Tremadoc at Chesley Drive. Abbreviation: TREM, Tremadoc. For explanation of symbols, see Figure 6. Modified from Landing (1980, fig. 3).

Examination of Matthew's topotypes of *Cyclognathus rotundifrons* indicates that they occur in sideritic, calcareous nodules identical to those in the lower 26 m of the Reversing Falls section (Figure 26; see Landing, 1980, fig. 3). Graptolites collected by EL from a horizon 25 m above the base of the Reversing Falls section include *Hunnegraptus* sp. cf. *H. copiosus* Lindholm and an *Adelograptus* with proximal thecae similar to those of *A. quasimodo* Rushton (Landing et al., 1997). These data mean that this section in the Chesley Drive Group (Landing, 1996a) is referable to the late Tremadoc, not Arenig, and a correlation with the Hunnebergian Stage is most appropriate. The Reversing Falls section and a *Peltocare rotundifrons*-bearing section on McLeod Brook, Cape Breton Island (Hutchinson, 1952), which has yielded a U-Pb zircon date of 483 ± 1 Ma on a K-bentonite (Landing et al., 1997), are the only presently known Hunnebergian Age sections in North American Avalon.

Chesley Drive

Chesley Drive: location.—A low ridge of overturned, steeply northwest-dipping black shale (Figure 27) is exposed in a roadcut on the north side of Rte. 100 (Chesley Drive) opposite the Ocean Steel Corp. plant. Vehicles should turn left (north) onto the first street after crossing the Reversing Falls bridge and park. The Chesley Drive cut lies 15—68 m west of the railroad spur that crosses Chesley Drive into the Ocean Steel storage yard.

Chesley Drive: Merionethian–Tremadoc.—This short section spans the Cambrian–Ordovician boundary, and is the type section for the dark grey shale-dominated, middle Upper Cambrian–upper Tremadoc Chesley Drive Group in Avalonian North America (Landing, 1996a). Earlier attempts to subdivide this stratigraphically long interval (i.e., depositional sequence 10) into formation-level units in eastern Newfoundland, Cape Breton Island, and the Saint John region have resulted either in biostratigraphically defined formations or units not differentiable on the basis of named lithologic criteria (Landing, 1996a). For these reasons, depositional sequence 10 has been assigned to a group-level unit, with the likelihood that formation-level units will be defined on the basis of regionally extensive stratigraphic markers. The thin Cambrian–Ordovician boundary sections at Chesley Drive and at Navy Island to the southeast under the Harbour Bridge (Rte. 1), and the upper Tremadoc section at Reversing Falls are reference sections for the Chesley Drive Group.

The Chesley Drive and Navy Island sections (Landing et al., 1978) have latest Cambrian and earliest Tremadoc faunas. Trilobite cranidia from a calcareous nodule (CD-9.2) are more similar to Late Cambrian than early Tremadoc species of the olenid genus *Parabolina* (M.E. Taylor *in* Landing, 1980). Dendroid graptolites from the west end of the cut include *Rhabdinopora flabelliforme parabola* (Bulman) and *R. flabelliforme sociale* (Salter) and are earliest Tremadoc. Hayes and Howell's (1937) stratigraphic scheme requires division of the black shale at this locality into an Upper Cambrian "Narrows Formation" and a Tremadoc "Navy Island Formation," or, according to Tanoli and Pickerill (1988), into a lower "Silver Falls Formation" and upper "Reversing Falls Formation" (all of the stratigraphic names in quotation marks were abandoned by Landing, 1996a).

Exit 113

Exit 113, location.—The south-dipping Exit 113 section (Figure 28) is exposed in a long roadcut on the west-bound lane of the Exit 113 interchange from Rte. 1. Vehicles can be parked on the wide margin of this exit ramp or in the neighborhood to the west. Landing and Westrop (1996) discussed the stratigraphic succession and depositional environments of the Exit 113 cut.

Exit 113: Placentian Series.—The Ratcliffe Brook Group seen at Somerset Street is faulted out of the Exit 113 section. Only the uppermost 0.8–1.0 m of medium-grained, light green to white quartz arenite of the middle Placentian Random Formation (depositional sequence 2) is present in fault contact with Coldbrook Group volcanics.

Exit 113: Branchian Series.—The Random Formation is overlain by 1.2 m of coarse-grained, pyritiferous quartz arenite with black phosphatic cement and scattered phosphatic pebbles. This feather edge of the St. Martins Member of the lower Hanford Brook Formation has yielded acrotretid brachiopods known higher in the Hanford Brook Formation. Neither acrotretids nor any other inarticulate brachiopods are known as low as the Random Formation, and this "black sandstone", which was earlier grouped with the underlying white sandstone in the "Glen Falls Formation" (Hayes and Howell, 1927), is now assigned to the Hanford Brook Formation and the lower part of depositional sequence 4. The Exit 113 section shows the cryptic unconformity between the middle Placentian and Branchian Series.

The overlying strata of the Hanford Brook Formation are referable to the Somerset Street Member, and the highest member, the Long Island Member, is absent. A 10 cm-thick black, phosphatic shale at the base of the Somerset Street Member has a large (4 cm-wide), looping *Didymaulichnus* trace. Medium-grey, fine-grained sandstone with abundant brachiopods and trilobites and larger traces alternate with darker grey, laminated, more pyritiferous, fine-grained sandstone with millimeter-sized burrows that record less and more strongly dysaerobic facies, respectively, in the low-energy Somerset Street Member.

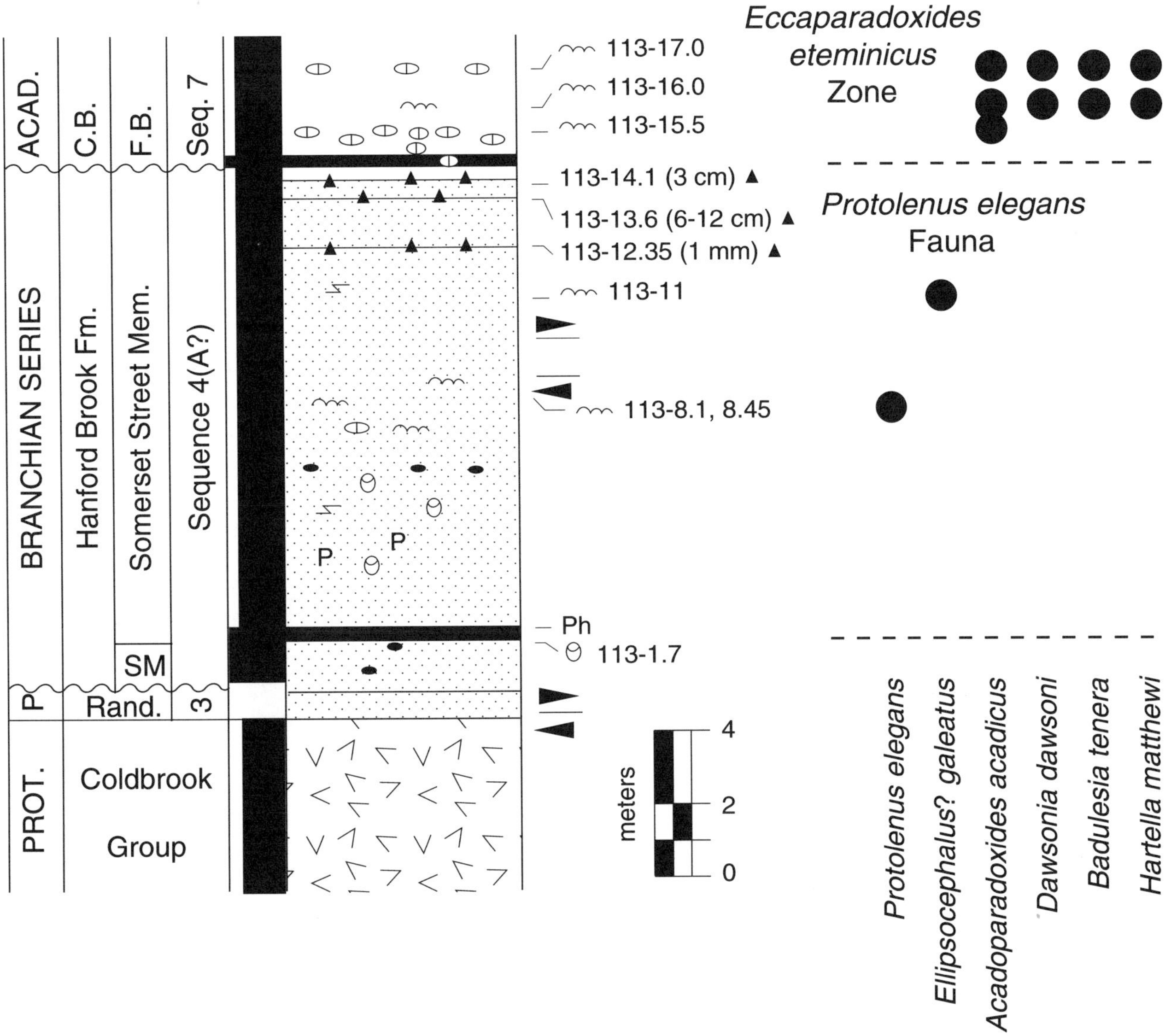

Figure 28. Proterozoic and middle Placentian–lower Acadian at Exit 113. For explanation of symbols, see Figure 6. Modified from Landing and Westrop (1996, fig. 3).

Three feldspar-rich volcanic ashes up to 12 cm-thick are present in the upper part of the Somerset Street Member, and two bedding plane-parallel faults (Figure 28) may have developed on ash horizons. The ash at 13.6 m just below the unconformable contact with the Middle Cambrian has zircons with a U-Pb age of 511 ± 1 Ma (Landing et al., In press).

Sparse trilobites of the *Protolenus elegans* Fauna occur in the middle Somerset Street Member at Exit 113 (Figure 28). The faunas are comparable to those of the Somerset Street Member at Hanford Brook West (August 19).

Exit 113: Acadian Series.—The Lower–Middle Cambrian boundary unconformity is marked by a 30 cm-thick, black, phosphatic quartz arenite with phosphatic pebbles (14.4-14.7 m). Overlying nodular trilobite packstones (14.7–15.5 m) were collected in the road ditch and contain wave-comminuted paradoxidiid fragments with rare helcionellids and ostracodes. These packstones are overlain by medium grey shale (15.5–17.9 m) with trilobites and 10 cm-thick, trilobite wackestone nodules at 17.0 m.

An *Eccaparadoxides eteminicus* Zone fauna just above the phosphatic sandstone lies in the lower Fossil Brook Member of the Chamberlain's Brook Formation. The contact between the Somerset Street Member and Fossil Brook Member marks the depositional sequence 4–7 unconformity between the Lower and Middle Cambrian. The absence of the Long Island Member of the upper Hanford Brook Formation (see Hanford Brook West stop on August 19) at Exit 113 and all localities in Saint John may be due to deeper erosion on an uplifted block in this part of New Brunswick (Landing and Westrop, 1996a).

Acknowledgments

Field and laboratory work for this synthesis was done under National Science Foundation grant EAR94-15773 (E. Landing) and Natural Sciences and Engineering Research Council of Canada grant 41197 (S.R. Westrop). G. Geyer and A.R. Palmer reviewed the manuscript.

References

Alcock, F.J. 1938. Geology of Saint John region, New Brunswick. Geological Survey of Canada Memoir 216, 65 p.

Anderson, M.M. 1987. Stratigraphy of Cambrian rocks at Bacon Cove, Duffs, and Manuels River, Conception Bay, Avalon Peninsula, eastern Newfoundland, p. 467–471. *In* D.C. Roy (ed.), Northeastern Section of the Geological Society of America, Centennial Field Trip Volume 5. Geological Society of America, 481 p.

————, and A.F. King. 1982. Manuels River Cambrian succession, p. 48–53. *In* A.F. King (ed.), The Caledonide Orogen, I.G.C.P. Project 27. NATO Advanced Study Institute, Publication No. 50.

Bengtson, S., and T.P. Fletcher. 1983. The oldest sequence of skeletal fossils in the Lower Cambrian of southeastern Newfoundland. Canadian Journal of Earth Sciences, 20:525–526.

Benus, A.P. 1988. Sedimentological context of a deep-water Ediacaran fauna (Mistaken Point Formation, Avalon Zone, eastern Newfoundland, p. 8, 9. *In* E. Landing, G.M. Narbonne, and P. Myrow (eds.), Trace Fossils, Small Shelly Fossils, and the Precambrian–Cambrian Boundary. New York State Museum Bulletin 463.

Bergstrom, J., and R. Levi–Setti. 1978. Phenotypic variation in the Middle Cambrian trilobite *Paradoxides davidis* Salter at Manuels, SE Newfoundland. Geologica et Palaeontologica, 12:1–40.

Brasier, M.D. 1984. Microfossils and small shelly fossils from the Lower Cambrian *Hyolithes* Limestone at Nuneaton, English Midlands. Geological Magazine, 121:229–253.

————. 1989. Sections in England and their correlation, p. 82–104. *In* J.W. Cowie and M.D. Brasier (eds.), The Precambrian–Cambrian Boundary. Oxford Monographs on Geology and Geophysics, No. 12.

Conway Morris, S. 1988. Radiometric dating of the Precambrian–Cambrian boundary in the Avalon Zone, p. 53—58. *In* E. Landing, G.M. Narbonne, and P. Myrow (eds.), Trace Fossils, Small Shelly Fossils, and the Precambrian-Cambrian Boundary. New York State Museum Bulletin 463.

————, and A.W.A. Rushton. 1988. Precambrian to Tremadoc biotas in the Caledonides, p. 93–109. *In* A.L. Harris and D.J. Fettes (eds.), The Caledonian-Appalachian Orogen. Geological Society of London Special Publication 38, 643 p.

Cowie, J.W., A.W.A. Rushton, and C.J. Stubblefield. 1972. A correlation of Cambrian rocks in the British Isles. Geological Society of London, Special Report 2, 42 p.

Currie, K.L. 1987. The Avalonian terrane around Saint John, New Brunswick, and its deformed Carboniferous cover, p. 403–408. *In* D.C. Roy (ed.), Centennian Field Guidebook, Volume 5, Northeastern Section of the Geological Society of America, 481 p.

Dale, N.C. 1915. The Cambrian manganese deposits of Conception and Trinity Bays, Newfoundland. Princeton University Contributions to the Geology of Newfoundland, No. 2. Proceedings of the American Philosophical Society, 54:371–456.

Dallmeyer, R.D., A.L. Odum, C.F. O'Driscoll, and E.M. Hussey. 1981. Geochronology of the Swift Current granite and host rocks of the Love Cove Group, southwestern Avalon Zone, Newfoundland: Evidence of a late Proterozoic volcanic–subvolcanic association. Canadian Journal of Earth Sciences, 18:699–707.

Davidek, K., E. Landing, S.A. Bowring, S.R. Westrop, A.W.A. Rushton, R.A. Fortey, and J.M. Adrain. In press. New uppermost Cambrian U-Pb date from Avalonian Wales and age of the Cambrian–Ordovician boundary. Geological Magazine.

Dawson, J.W. 1868. Acadian Geology. MacMillan and Company, London, 694 p.

Ekdale, A.A., R.G. Bromley, and S.G. Pemberton. 1984. Ichnology--The use of trace fossils in sedimentology and stratigraphy. Society of Economic Paleontologists and Mineralogists, Short Course No. 15, 317 p.

Fernsides, W.G. 1910. The Tremadoc slates and associated rocks of southeast Carnarvonshire. Quarterly Journal of the Geological Society of London, 66:142–188.

Fletcher, T.P. 1972. Geology and Lower to Middle Cambrian trilobite faunas of the southwest Avalon, Newfoundland. Unpub. Ph.D. dissertation, University of Cambridge, 520 p.

Geyer, G., and A.R. Palmer. 1995. Nelteriidae and Holmiidae (Trilobita) from Morocco and the problem of Early Cambrian intercontinental correlation. Journal of Paleontology, 69:459–474.

————, and A. Uchman. 1995. Ichnofossil assemblages from the Nama Group (Neoproterozoic–Lower Cambrian) in Namibia and the Proterozoic–Cambrian boundary problem reconsidered, p. 175–202. *In* G. Geyer and E. Landing (eds.), Morocco '95—The Lower–Middle Cambrian Standard of Western Gondwana. Beringeria Special Issue 2.

Hayes, A.O., and B.F. Howell. 1937. Geology of Saint John. Geological Society of America, Special Paper 5, 146 p.

Helmstead, H., and S. Tella 1972. Structural history of pre-Carboniferous rocks in parts of eastern Cape Breton Island. Report of Activities, Part A, Geological Survey of Canada, Paper 72-1:7–10.

————, and ————. 1973. Pre-Carboniferous structural history of southeast Cape Breton Island, Nova Scotia. Maritime Sediments, 9:88–99.

Henningsmoen, G. 1957. The trilobite family Olenidae. Skrifter utgitt av det Norske Videnskaps-Akademi i Oslo 1, Matematisk-Naturvidenskapelig Klasse 1, 303 p.

Hofmann, H.J., and I.M. Patel 1989. Trace fossils from the type `Etcheminian Series' (Lower Cambrian Ratcliffe Brook Formation), Saint John area, New Brunswick, Canada. Geological Magazine, 126:139–157.

Howell, B.F. 1925. The faunas of the Cambrian *Paradoxides* beds at Manuels, Newfoundland. Bulletins of American Paleontology, 11(43), 140 p.

Hughes, C.J., and W.D. Bruckner. 1972. Late Precambrian rocks of the eastern Avalon Peninsula, Newfoundland—a volcanic island complex. Canadian Journal of Earth Sciences, 8:899–915.

Hutchinson, R.D. 1952. The stratigraphy and trilobite faunas of the Cambrian sedimentary rocks of Cape Breton Island, Nova Scotia. Geological Survey of Canada Memoir 263, 124 p.

————. 1953. Geology of Harbour Grace Map-Area, Newfoundland. Geological Survey of Canada Memoir 275, 43 p.

———. 1962. Cambrian stratigraphy and trilobite faunas in southeast Newfoundland. Geological Survey of Canada Bulletin 88, 156 p.

Jenness, S.E. 1963. Terra Nova and Bonavista Map-Areas, Newfoundland (2D E1/2 and 2C). Geological Survey of Canada, Memoir 327, 184 p.

King, L.H., G.B.J. Fader, W.H. Poole, and R.K. Wanless. 1985. Geological setting and age of the Flemish Cap granodiorite, east of the Grand Banks of Newfoundland. Canadian Journal of Earth Sciences, 22:1286–1298.

Krogh, T.E., D.F. Strong, and V. Papezik. 1983. Precise U-Pb ages of zircons from volcanic and plutonic units in the Avalon Peninsula. Geological Society of America, Abstracts and Programs, 15:136.

Landing, E. 1980. Late Cambrian-Early Ordovician macrofaunas and phosphatic microfaunas, St. [sic, read Saint] John Group, New Brunswick. Journal of Paleontology, 54:752–761.

———. 1991. Upper Precambrian through Lower Cambrian of Cape Breton Island: faunas, paleoenvironments, and stratigraphic revision. Journal of Paleontology, 65:570–595.

———. 1992. Lower Cambrian of southeastern Newfoundland: epeirogeny and Lazarus faunas, lithofacies-biofacies linkages, and the myth of a global chronostratigraphy, p. 283–309. *In* J.H. Lipps and P.W. Signor (eds.), Origin and Early Evolution of the Metazoa. Plenum Press, New York.

———. 1994. Precambrian—Cambrian boundary ratified and a new perspective of Cambrian time. Geology, 22:179–182.

———. 1995. Upper Placentian–Branchian Series of mainland Nova Scotia (middle–upper Lower Cambrian): faunas, paleoenvironments, and stratigraphic revision. Journal of Paleontology, 69:475–496.

———. 1996a. Avalon: insular continent by the latest Precambrian, p. 29–63. *In* R.D. Nance and M.D. Thompson (eds.), Avalonian and Related Peri-Gondwanan Terranes of the Circum-North Atlantic. Geological Society of America Special Paper 304.

———. 1996b. Reconstructing the Avalon continent: marginal to inner platform transition in the Lower Cambrian of southern New Brunswick. Canadian Journal of Earth Sciences, 33:1185–1192.

———, and A.P. Benus. 1988a. Cambrian depositional history and stratigraphy: Avalon–Bonavista region, southeastern Newfoundland. Field Trip Guidebook, Trip A3, Geological Association of Canada, St. John's, 30 p.

———, and ———. 1988b. Stratigraphy of the Bonavista Group, southeastern Newfoundland: growth faults and the distribution of the subtrilobitic Lower Cambrian, p. 59–71. *In* E. Landing, G.M. Narbonne, and P. Myrow (eds.), Trace Fossils, Small Shelly Fossils, and the Precambrian–Cambrian Boundary. New York State Museum Bulletin 463.

———, and C.E. Brett 1987. Trace fossils and regional significance of a Middle Devonian (Givetian) disconformity in southeastern Ontario. Journal of Paleontology, 61:205–230.

———, and R.F. Miller. 1988. Bibliography of George Frederic Matthew, p. 77–80. *In* E. Landing, G.M. Narbonne, and P. Myrow (eds.), Trace Fossils, Small Shelly Fossils, and the Precambrian—Cambrian boundary. New York State Museum Bulletin 463, 81 p.

———, and J.B. Murphy. 1991. Uppermost Precambrian(?)–Lower Cambrian of mainland Nova Scotia: faunas, depositional environments, and stratigraphic revision. Journal of Paleontology, 65:382–396.

———, and S.R. Westrop. 1996. Upper Lower Cambrian depositional sequence in Avalonian New Brunswick. Canadian Journal of Earth Sciences, 33:404–417.

———, M.E. Taylor, and B.-D. Erdtmann. 1978. Correlation of the Cambrian–Ordovician boundary between the Acado-Baltic and North American faunal provinces. Geology, 6:75–78.

———, G.M. Narbonne, P. Myrow A.P. Benus, and M.M. andErson. 1988. Faunas and depositional environments of the upper Precambrian through Lower Cambrian, southeastern Newfoundland, p. 18–52. *In* E. Landing, G.M. Narbonne, and P. Myrow (eds.), Trace Fossils, Small Shelly Fossils, and the Precambrian-Cambrian Boundary. New York State Museum Bulletin 463, 81 p.

———, P. Myrow, A.P. Benus, and G.M. Narbonne. 1989. The Placentian Series: appearance of the oldest skeletalized faunas in southeastern Newfoundland. Journal of Paleontology, 63:739–769.

———, S.A. Bowring, R.A. Fortey, K.L. Davidek. 1997. U-Pb zircon date from Avalonian Cape Breton Island and geochronologic calibration of the Early Ordovician. Canadian Journal of Earth Sciences, 34:724–730.

———, ———, K. Davidek, S.R. Westrop, G. Geyer, and W. Heldmaier. In press. New U-Pb zircon dates from Avalon and Gondwana and duration of the Early Cambrian. Canadian Journal of Earth Sciences.

McCartney, W.D. 1967. Whitbourne Map-Area, Newfoundland. Geological Survey of Canada Memoir 341, 135 p.

———, W.H. Poole, R.K. Wanless, H. Williams, and W.D. Loveridge. 1966. Rb/Sr age and geological setting of the Holyrood granite, southeastern Newfoundland. Canadian Journal of Earth Sciences, 3:947–957.

McCutcheon, S.R. 1987. Cambrian stratigraphy in the Hanford Brook area, southern New Brunswick, Canada, p. 399–402. *In* D.K. Roy (ed.), Centennial Field Trip Guide Volume 5, Northeastern Section of the Geological Society of America, 481 p.

McLearn, F.H. 1915. The Lower Ordovician (*Tetragraptus* Zone) at St. John, New Brunswick, and the new genus *Protistograptus*. American Journal of Science, 40:49–59.

Martin, F., and W.T. Dean. 1981. Middle and Upper Cambrian and Lower Ordovician acritarchs from Random Island, eastern Newfoundland. Geological Survey of Canada Bulletin 343.

———, and ———. 1988. Middle and Upper Cambrian acritarch and trilobite zonation at Manuels River and Random Island, eastern Newfoundland. Geological Survey of Canada, Bulletin 381.

Matthew, G.F. 1863. Observations on the geology of Saint John County, New Brunswick. Canadian Naturalist, 8:241–259.

———. 1885. Illustrations of the fauna of the St. John Group continued, No. III; descriptions of new genera and species (including a description of a new species of *Solenopleura* by J.F. Whiteaves). Transactions of the Royal Society of Canada, 3(4):29–84 (issued in 1886).

———. 1889. On Cambrian organisms in Acadia. Transactions of the Royal Society of Canada, 7(4):135–162.

———. 1892. Illustrations of the fauna of the St. John Group. No. VII. Transactions of the Royal Society of Canada, 10(4):95–109.

———. 1895. The *Protolenus* fauna. Transactions of the New York Academy of Sciences, 14:101–153.

———. 1896. Faunas of the *Paradoxides* beds in eastern North America. Transactions of the New York Academy of Sciences, 15:192-197.

———. 1899. A Palaeozoic terrane beneath the Cambrian. Annals of the New York Academy of Sciences, 12:41–56.

Miller, R.F. 1988. George Frederic Matthew (1837–1923), p. 4–7. *In* E. Landing, G.M. Narbonne, and P. Myrow (eds.), Trace Fossils, Small Shelly Fossils, and the Precambrian–Cambrian Boundary. New York State Museum Bulletin 463, 81 p.

Myrow, P.M., and R.N. Hiscott. 1993. Depositional history and sequence stratigraphy of the Precambrian–Cambrian global stratotype section, Chapel Island Formation, southeast Newfoundland. Palaeogeography, Palaeoclimatology, Palaeoecology, 104:13–35.

———, and E. Landing. Mixed siliciclastic-carbonate deposition in an Early Cambrian oxygen-stratified basin, Chapel Island Formation, southeastern Newfoundland. Journal of Sedimentary Petrology, 62:455–473.

———, G.M. Narbonne, and R.N. Hiscott. 1988. Storm-shelf and tidal deposits of the Chapel Island and Random Formations, Burin Peninsula: facies and trace fossils. Field Trip Guidebook, Trip B6. Geological Association of Canada, St. John's, 50 p.

Narbonne, G.M., and P.M. Myrow. 1988. Trace fossil biostratigraphy in the Precambrian—Cambrian boundary interval, p. 72–76. *In* E. Landing, G.M. Narbonne, and P.M. Myrow (eds.), Trace Fossils, Small Shelly Fossils, and the Precambrian-Cambrian Boundary. New York State Museum Bulletin 463, 81 p.

———, ———, E. Landing, and M.M. Anderson. 1987. A candidate stratotype for the Precambrian—Cambrian boundary, Fortune Head, Burin Peninsula, southeastern Newfoundland. Canadian Journal of Earth Sciences, 24:1277–1293.

———, ———, ———, ———. 1991. A chondrophorine (medusoid hydrozoan) from the basal Cambrian (Placentian) of Newfoundland. Journal of Paleontology, 65:186–191.

North American Commission on Stratigraphic Nomenclature. 1983. North American Stratigraphic Code. American Association of Petroleum Geologists Bulletin, 67:841–875.

Palmer, A.R. 1969. Cambrian trilobite distributions in North America and their bearing on Cambrian paleogeography of Newfoundland, p. 139–144. *In* M. Kay (ed.), North Atlantic–Geology and Continental Drift. American Association of Petroleum Geologists, Memoir 12, 1082 p.

Patel, I.M. 1977. Late Precambrian fossil soil horizon in southern New Brunswick, and its stratigraphic significance. Geological Society of America, Abstracts with Programs, 9:308.

———. 1985. Trace fossils from an early Lower Cambrian shallow water sequence from southern New Brunswick. Canadian Paleontology and Biostratigraphy Seminar, Program with Abstracts, Universite Laval, p. 32.

Pe-Piper, G., D.J.W. Piper, C.N. Kotopouli, and A.G. Panagos. 1995. Neogene volcanoes of Chios-Greece: the relative importance of subduction and back-arc-extension, p. 213–231. *In* J.L. Smellie (ed.), Volcanism Associated with Extension at Consuming Plate Margins. Geological Society Special Publication 81.

Poulsen, V., and M.M. Anderson. 1975. The Middle–Upper Cambrian transition in southeastern Newfoundland, Canada. Canadian Journal of Earth Sciences, 12:2065–2079.

Rast, N., B.H. O'Brien, and R.J. Wardle. 1976. Relationships between Precambrian and Lower Palaeozoic rocks of the 'Avalon Platform' in New Brunswick, the northeast Appalachians and the British Isles. Tectonophysics, 30:315–338.

Robison R.A. 1984. Cambrian Agnostida of North American and Greenland. Part 1, Ptychagnostidae. University of Kansas Paleontological Contributions, Paper 109, 59 p.

———. 1994. Agnostoid trilobites from the Henson Gletcher and Kap Stanton Formations (Middle Cambrian), North Greenland. Groenlands Geologiske Undersoegelse, Bulletin 169:25–77

———. 1995. Revision of the Middle Cambrian trilobite *Agnostus acadicus* Hartt. Journal of Paleontology, 69:302–307.

———, A.V. Rosova, A.J. Rowell, and T.P. Fletcher. 1977. Cambrian boundaries and divisions. Lethaia, 10:257–262.

Rushton, A.W.A. 1974. The Cambrian of Wales and England, p. 43–121. *In* C.H. Holland (ed.), Cambrian of the British Isles, Norden, and Spitsbergen. John Wiley and Son, New York.

———. 1978. Fossils from the Middle–Upper Cambrian transition in the Nuneaton district. Palaeontology, 21:245–283.

————. 1982. The biostratigraphy and correlation of the Merioneth–Tremadoc Series boundary in north Wales, p. 41–59. *In* M.G. Bassett and W.T. Dean (eds.), The Cambrian–Ordovician Boundary: Sections, Fossil Distributions, and Correlations. National Museum of Wales, Geological Series No. 3, 227 p.

Sepkoski, J.J., JR. 1991. A model of onshore–offshore change in faunal diversity. Paleobiology, 17:58–77.

Smith, S.A., and R.N. Hiscott. 1984. Latest Precambrian to Early Cambrian basin evolution, Fortune Bay, Newfoundland: fault-bounded basin to platform. Canadian Journal of Earth Sciences, 21:1379–1392.

Steel, R.J., and T.G. Gloppen. 1980. Late Caledonian (Devonian) basin formation, western Norway: signs of strike-slip tectonics during basin infilling, p. 7–26. *In* H.F. Balance and H.G. Reading (eds.), Sedimentation in Oblique-Slip Mobile Belts. International Association of Sedimentologists, Special Publication 4.

Tanoli, S.K., and R.K. Pickerill. 1988. Lithostratigraphy of the Cambrian–Lower Ordovician Saint John Group, southern New Brunswick. Canadian Journal of Earth Sciences, 25:669–690.

Taylor, K., and A.W.A. Rushton. 1971. The pre-Westphalian geology of the Warwickshire Coalfield with a description of three boreholes in the Merevale area. Bulletin of the Geological Survey of Great Britain, No. 35, 150 p.

Theokritoff, G. 1979. Early Cambrian biogeography in the North Atlantic region. Lethaia, 18:283–293.

Van Ingen, G. 1914. Table of geological formations of the Cambrian and Ordovician Systems about Conception and Trinity Bays, Newfoundland. Princeton University Contributions to the Geology of Newfoundland, 4, 1 p.

Van Wagoner, J.C., H.W. Postamentier, R.M. Mitchum, P.R. Vail, J.F. Sarg, T.S. Louit, and J. Hardenbol. 1988. An overview of the fundamentals of sequence stratigraphy and key definitions, p. 39–45. *In* C.K. Wilgus, H.W. Postamentier, C.A. Ross, and C.G. St. C. Kendall (eds.), Sea-Level Changes: An Integrated Approach. Society of Economic Paleontologists and Mineralogists, Special Publication 42, 407 p.

Walcott, C.D. 1900a. Random, a Precambrian upper Algonkian terrane. Bulletin of the Geological Society of America, 11:3–5.

————. 1900b. Lower Cambrian terrane in the Atlantic Province. Proceedings of the Washington Academy of Science, 1:301–399.

Westergaard, A.H. 1946. Agnostoidea of the Middle Cambrian of Sweden. Sveriges Geologiska Undersokning, Series C, No. 477, 140 p.

Westrop, S.R., J.V. Trembley, and E. Landing. 1995. Declining importance of trilobites in Ordovician nearshore communities: dilution or displacement? Palaios, 10:75–79.

Williams, G.L., L.R. Fyffe, R.J. Wardle, S.P. Colman-Sadd, R.C. Boehner, J.A. Watt, J. Bryant, M.D. Daneau, W.C. MacMillan, K.M. Reardon, and D.E. Webber. 1985. Lexicon of Canadian Stratigraphy, Volume VI. Canadian Society of Petroleum Geologists, Calgary, 572 p.

Young, T., F. Martin, W.T. Dean, and A.W.A. Rushton. 1994. Cambrian stratigraphy of St. Tudwal's Peninsula, Gwynnedd, northwest Wales. Geological Magazine, 131:335–360.

Revisions in Stratigraphic Nomenclature of the Cambrian of Avalonian North America and Comparisons with Avalonian Britain

Ed Landing and Stephen R. Westrop[*]

New York State Geological Survey, The State Education Department, Albany, NY 12230, and Department of Earth Sciences, Brock University,St. Catharines, Ontario L2S 3A1
[*]*Current address: Oklahoma Museum of Natural History and School of Geology and Geophysics, University of Oklahoma, Norman, Oklahoma 73019*

Abstract

Examination of the stratigraphy of Avalonian North America has led to recognition of several formation- and member-level units that reflect Cambrian sequence-stratigraphic developments. The upper Lower Cambrian Brigus Formation is composed of two siliciclastic mudstone-dominated depositional sequences in southeast Newfoundland. These are the St. Mary's Member (new, sequence 4A) and the upper, volcanic ash-bearing Jigging Cove Member (new, depositional sequence 4B). The St. Mary's–Jigging Cove contact corresponds to a faunal transition between the *Callavia broeggeri* Zone and *Myopsostrenua?* cf. *M.? sabulosa* Faunal Interval. This faunal and lithologic boundary may be present in siliciclastic mudstone-dominated Branchian Series successions in Avalonian Britain. Upper Lower Cambrian (Branchian Series) strata of southern New Brunswick consist of the Hanford Brook Formation. The base of the representative section of the Long Island Member of the upper Hanford Brook Formation on Hanford Brook is revised downward to the base of the massive, coarse-grained sandstone bed, which marks a sequence boundary. This redefined contact of the Somerset Street and Long Island members may correspond to the depositional sequence 4A–4B contact in southeastern Newfoundland and may mean that the *Protolenus elegans* Fauna of the Somerset Street Member is a sandstone lithofacies equivalent of the *C. broeggeri* Zone in siliciclastic mudstones.

The upper Middle Cambrian Manuels River Formation (depositional sequence 8) is composed of four alternations of weakly and strongly dysaerobic siliciclastic mudstones. These alternations are recognized across southeastern Newfoundland as members 1–4 (new) of the Manuels River Formation. Late Middle Cambrian basin reorganization marked by basalt volcanism late during Manuels River Formation deposition (Hay Cove volcanics) and early during depositional sequence 9 accumulation (i.e., Chapel Arm Member volcanics) is shown by erosion of member 4 in areas outside of the St. Mary's–east Trinity axis. This erosion was followed by onlap and deposition of the lenticular lithosome sandstone-dominated Cavendish Formation (new, locally to 45 m thick) at the base of depositional sequence 9. The Cavendish Formation, earlier included in the top of the Manuels River Formation (revised), is the basal unit of the MacLean Brook Group (revised). The uppermost Middle Cambrian depositional sequence 8–9 unconformity between the Manuels River and Cavendish Formations is lithologically identical to and coeval with the Nant-y-big Formation–"*Dorypyge* Limestone" and Abbey Shales–Mancetter Formation unconformities in Avalonian Britain.

Introduction

Work since a recent revision of the uppermost Proterozoic–Lower Ordovician stratigraphic nomenclature of Avalonian North America (Landing, 1996) has led to the recognition of a number of new formation- and member-level units that require formal designation. These units record important developments in the sequence stratigraphy of the Avalonian cover sequence. In addition, minor revisions involving changes in boundaries and rank are required for several existing stratigraphic units.

Because the stratigraphic context of these new and revised units has been discussed and illustrated in the field trip guide in this volume, only relatively brief definitions and discussions are required for these units. Locations and dates in the field trip guide are referred to below in the designation of type and reference sections and in the discussion of the stratigraphic nomenclature.

Brigus Formation (New Members)

St. Mary's Member (new).—The St. Mary's Member is proposed as the lowest member of the Brigus Formation and trilobite-bearing, upper Lower Cambrian (i.e., Branchian Series) in southeastern Newfoundland and other areas in North American Avalon. The member is a siliciclastic mudstone-dominated unit with: 1) red, purple, and green colors; 2) unconformable lower and upper contacts that correspond to depositional sequence boundaries; 3) variable amounts (i.e., absent to locally important) of bedded and nodular limestones; and 4) generally, but not invariably, minor development of sandstones. Recorded thicknesses of the St. Mary's Member range from 12 m near the eastern feather-edge in the southern Burin Peninsula (August 14, Little Dantzig Cove Brook) to 118.4 m at Long Cove in the St. Mary's–east Trinity axis (Figure 2, locality LC; E. Landing, unpub. data).

The type section of the St. Mary's Member is the 86.65 m-thick section through the lower Brigus Formation on the northwest shore of Jigging Cove, north of the mouth of Red Head Brook (Figure 15). The name "St. Mary's" is unoccupied for Cambrian–Ordovician units in Avalon, and is applied as a member name to refer to the location of the type section on the west side of St. Mary's Bay.

Key to the definition of the St. Mary's Member are its unconformable lower and upper contacts. The member unconformably overlies strata as high as the uppermost Placentian Fosters Point Formation in the Placentia– Bonavista axis and parts of the St. Mary's–east Trinity axis in southeastern Newfoundland (August 15, Jigging Cove; August 16, Smith Point), but unconformable relationships with the lower Placentian Random Formation (August 14, Little Dantzig Cove Brook) or any older unit are possible. In east Conception Bay, the St. Mary's Member locally overlies rocks of the late Proterozoic Avalonian orogen (August 12, Red Bridge Road), and boulder deposits that accumulated at the base of a sea cliff may be referable to the member's base (August 12, Manuels River overlook).

The lowest unit of the St. Mary's Member is often a thin (0.25–2.0 m) trilobite wacke- to packstone or locally a linguloid-bearing fossil wackestone with exceptionally rare trilobites (August 13 and 14, Duffs and Chapel Cove) (Landing and Benus, 1988, figs. 35, 36). These limestones were earlier included in the top of the "Smith Point Limestone" (designation abandoned by Landing and Benus, 1988) in the earlier literature, but a phosphatized erosion surface (August 12, Duffs) or a bluish-weathering incipient soil horizon at the top of the sub-trilobitic Fosters Point Formation is often evident under the basal trilobite-bearing limestone of the St. Mary's Member (August 16, Smith Point section). In sections in southern and eastern Trinity Bay, lower St. Mary's Member strata immediately above the basal limestone are red, burrow-mottled to -churned, fine-grained sandstones that may reach 16 m in thickness (e.g., Long Cove section, Figure 2, locality LC; E. Landing, unpublished data). These sandstones seem to reflect the accumulation of sands that were eroded during an interval following sub-aerial exposure of the Placentian Series and transported into the St. Mary's–east Trinity axis.

The unconformable upper contact of the St.

Mary's Member is a depositional sequence boundary that marks a greater or lesser hiatus. When the upper sequence boundary falls within the Brigus Formation, the top of the St. Mary's Member may preserve the lithologies characteristic of the cap of an Avalonian shale basin shoaling cycle (i.e., red siliciclastic mudstone with a thin interval [ca. 1–2 m] of nodular limestones and an uppermost fossiliferous limestone; see Landing and Benus, 1988). In this case, the unconformity is marked by immediately overlying remane sediment-containing rocks that form the base of the Jigging Cove Member (new). These rocks at the base of the Jigging Cove Member include: 1) a thin, oncoid-bearing limestone or fine-grained, glauconitic sandstone that contains phosphatic, shale, and rare Proterozoic volcanic clasts (August 15, Jigging Cove, Figure 15; August 16, Smith Point, Figure 16) or 2) a phosphate pebble bed with siltstone matrix and locally abundant trilobites and lingulelloids. This phosphate pebble bed is found at the following localities: 1) Redlands Point, south end of St. Mary's Peninsula, 58 m above base of Brigus Formation (E. Landing and S.R. Westrop, unpub. data); 2) Long Cove, southern Trinity Bay, 118.4 m above base of Brigus Formation (E. Landing and S.R. Westrop, unpub. data); and 3) Hopeall Head, eastern Trinity Bay, 20 cm bed at 80.5-80.7 above base of Brigus Formation (Hutchinson, 1962, collection 20403; E. Landing and S.R. Westrop, unpub. data).

The St. Mary's Member is geographically widespread in Avalonian North America. It is lithologically comparable to the *Callavia broeggeri* Zone assemblage-bearing Brigus Formation exposed in the northern Antigonish Highlands, mainland Nova Scotia (Landing, 1995). In eastern Massachusetts, the member is recognizeable above the Fosters Point Formation (i.e., "Nahant Limestone") at Nahant Point, northeast of Boston, and composes the trilobite-bearing part of the Hoppin Reservoir section (see Landing, 1988).

In a number of regions of southeastern Newfoundland, the St. Mary's Member comprises the entire Brigus Formation. Absence of the overlying Jigging Cove Member may be explained by: 1) erosion prior to deposition of the Middle Cambrian Chamberlain's Brook Formation that removed the Jigging Cove Member, or 2) the Jigging Cove Member may not have been deposited at great distances east and west of the St. Mary's–east Trinity axis. For example, in the southern Burin Peninsula (August 14, Little Dantzig Cove) and locally on the east side of Conception Bay (August 12, Red Bridge Road), the St. Mary's Member comprises the entire Brigus Formation, and the Jigging Cove Member is not present. However, the Jigging Cove Member appears above a remanie sediment bed just north of Red Bridge Road at the Manuels River bridge overlook (August 12).

The stratigraphic relationships of the St. Mary's Member (Figure 3) mean that it corresponds to the *Callavia broeggeri* Zone of Landing (1992). The unit overlies strata as high as the uppermost *Camenella baltica* Zone of the Fosters Point Formation, and trilobites of the *Myopsostrenua?* cf. *M.? sabulosa* Faunal Interval appear within the basal bed of the Jigging Cove Formation at such localities such as Smith Point, western Trinity Bay (August 16) and Redlands Point, St. Mary's Peninsula, (Figure 2, locality Re; S.R. Westrop and E. Landing, unpub. data). Because a number of *C. broeggeri* Zone taxa persist into the *"S." sabulosa* Zone (e.g., *Serrodiscus bellimarginatus, S. weymouthioides, Hebediscus attleborensis, Ladadiscus llarenai*; Fletcher, 1972), the St. Mary's–Jigging Cove hiatus may not represent a long time interval.

A regionally extensive marker bed associated with a faunal change immediately above the St. Mary's Member, and somewhat lithologically distinct facies of the overlying Jigging Cove Member (described below), make the St. Mary's Member a recognizable and mappable lithosome in southeastern Newfoundland. In an unpublished dissertation, Fletcher (1972) proposed a "Redland Cove Member" for the lower, calcareous nodule- and limestone-bearing part of the Brigus Formation. Even though proposal of a named stratigraphic unit in a dissertation does "not constitute publication within the meaning of the Code" (North American Commission on Stratigraphic Nomenclature, 1983, p. 851), a misnamed "Redlands Member" has been used in several published reports (e.g., Bengtson and

Fletcher, 1983; Brasier, 1989). The problem with the "Redlands Member" is that its upper limit is not recognizable away from Redlands, because the supposedly diagnostic calcareous nodules and bedded limestones of this lower part of the Brigus Formation persist to the very top of the formation.

Jigging Cove Member (new).—The Jigging Cove Member is the upper part of the Brigus Formation in southeast Newfoundland and other regions of Avalonian North America. The unit consists of red, purple, and green siliciclastic mudstones with minor nodular and very rare bedded limestones; a green-grey, pyritiferous, laminated siliciclastic mudstone with methanogenic carbonate nodules is commonly the lowest unit of the member. Thin volcanic ashes occur in the member in the southern St. Mary's–east Trinity axis (August 15, Branch and Jigging Coves) and western Trinity Bay (August 16, Smith Point). Its upper and lower contacts are sequence boundaries in southeast Newfoundland with the underlying St. Mary's Member of the Brigus Formation and with the overlying Chamberlain's Brook Formation. The type section of the Jigging Cove Member is the upper 63.45 m of the Brigus Formation above the type section of the St. Mary's Member on the northwest shore of Jigging Cove (August 15, Figure 15). The base of the member is exposed at Jigging Cove, and the contact with the overlying Chamberlain's Brook Formation is exposed at the reference section on the north shore of Branch Cove, St. Mary's Bay (August 15, Figure 14). The member is named for the Jigging Cove type locality.

Exposures of the Jigging Cove Member are limited to the Avalon Peninsula and western Trinity Bay areas in southeast Newfoundland, and the member is apparently cut out below the sub-Chamberlain's Brook unconformity in the southern Burin Peninsula and parts of eastern Conception Bay (discussed above). Thicknesses of the Jigging Cove Member range from 1.8 m at the eastern feather-edge of the member on Manuels River, eastern Conception Bay (August 12, Manuels River Bridge overlook), to 103+ m (base not exposed) at Branch Cove (Figure 14).

As with the St. Mary's Member, the sequence boundaries at the base and top of the Jigging Cove Member are important to its recognition because these two members of the Brigus Formation do not differ strongly in lithology. The oncoid-bearing limestone, glauconitic sandstone, or phosphate granule bed at the base of the Jigging Cove Member (and base of the *Myopsostrenua?* cf. *M.? sabulosa* Faunal Interval) has been discussed above under the definition of the St. Mary's Member. This remanie sediment-containing bed is typically overlain by green or green-grey, laminated siliciclastic mudstone that is weakly burrowed and lacks skeletal fossils (see Figures 15, 16). This dysaerobic mudstone facies marks a deepening into an oxygen-minimum zone on the Avalon shelf (compare Myrow and Landing, 1993) early in deposition of the Jigging Cove Member (depositional sequence 4B).

Siliciclastic mudstones of the upper Jigging Cove Member differ somewhat from those of the underlying St. Mary's Member. In general, the red mudstone facies of the Jigging Cove Member show a subtle brownish cast not present in the red mudstones of the St. Mary's Member. These brownish red mudstones are not as burrow-churned and structureless as the red mudstones of the St. Mary's Member and tend to retain such primary structures as bi-directional (wave) micro-cross-beds and trough cross-beds that seem to indicate the development of mud dunes. This somewhat restricted burrowing during deposition of the Jigging Cove Member may explain the abundance of thin volcanic ashes at the Jigging Cove and Branch Cove sections (i.e., ashes deposited during St. Mary's deposition may have been churned into underlying sediment and rendered unrecognizable). In addition, lenticular (to 1 m-thick), bedding plane-parallel, greenish "splotches" with purplish margins that are transitional into the surrounding red mudstone are common in the Jigging Cove Member and are rare in the St. Mary's Member. Comparable "splotchy" mudstones occur in the upper part of the poorly exposed Brigus Formation in Cape Breton Island (="Canoe Brook Formation" of Hutchinson, 1952), and the Jigging Cove mudstones are tentatively recognized in Cape Breton Island. Nodular (diagenetic) limestones and rare bedded limestones occur throughout the Jigging Cove Member, and are not distinguishable from those of the underlying St. Mary's Member.

Long sections of the Jigging Cove Member extend from the *Myopsostrenua?* cf. *M.? sabulosa* Faunal Interval at its base into Lower–Middle Cambrian boundary strata with *Oryctocara granulata* at the very top of the member (see August 15, Branch Cove). These data mean that the Jigging Cove Member corresponds to the very uppermost upper Lower Cambrian of Avalon.

The designations "Jigging Cove Mudstones Member" and "Branch Cove Marls Member" were used in a dissertation for the middle and upper parts of the Brigus Formation by Fletcher (1972). However, neither of these names were published "within the meaning of the Code" (North American Commission on Stratigraphic Nomenclature, 1983, p. 851), and the geographic locality "Jigging Cove" is available for use in naming a stratigraphic unit. The sequence boundary that separates the St. Mary's and Jigging Cove Members into mappable units has not been previously recognized, and falls within Fletcher's (1972) unpublished "Redlands Cove Member."

Comparison with Avalonian Britain.—Because depositional sequence boundaries are regionally extensive in Avalonian North America and southern Britain (Landing, 1996), lithologic developments comparable to those seen at the depositional sequence 4A–4B boundary interval in southeastern Newfoundland and, potentially, in southern New Brunswick (discussed below under Hanford Brook Formation) may appear in other Avalonian regions. For example, upper Branchian Series strata that are dominated by red and green siliciclastic mudstones range at least up into *Myopsostrenua?* cf. *M.? sabulosa* Faunal Interval-equivalent strata in north Wales (upper Llanberis Slates; Morris and Fernsides, 1928; Howell and Stubblefield, 1950) and through the Lower–Middle Cambrian boundary in Warwickshire, England (lower Purley Shales; Rushton, 1966). These regions of British Avalon may preserve a record of the sequence 4A-4B depositional history known in Avalonian North America. Interestingly, A.W.A. Rushton (personal commun., 1997) recollects the presence of a probable K-bentonite in a temporary excavation in red shales that yielded upper Lower Cambrian trilobites in the Camp Hill area, near Nuneaton, Warwickshire (Rushton, 1966, locality 2A). This occurrence and its age is reminiscent of the volcanic ash-bearing Jigging Cove Member in southeastern Newfoundland. The potential exists for a unified stratigraphic nomenclature for the Branchian Series on the inner platform of southeastern Newfoundland and Warwickshire (see Landing, 1996, figs. 1, 7), with the Brigus Formation (Van Ingen, 1914) as an appropriate synonym for the lower part of the Brigus—Chamberlain's Brook-equivalent Purley Shales of Lapworth (1898).

Hanford Brook Formation (Revised Members)

Landing and Westrop (1996) interpreted the upper Lower Cambrian (Branchian Series) Hanford Brook Formation in southern New Brunswick as a deepening–shoaling depositional sequence composed of three successive members. The deeper-water, dysaerobic, siltstone and fine-grained sandstone facies of the Somerset Street Member that form the middle of the formation are overlain by shallower, better oxygenated facies with locally prominent hummocky cross-stratification of the Long Island Member of the uppermost Hanford Brook Formation.

A succession through the Somerset Street and Long Island Members is exposed only on Hanford Brook (Figure 19, locality HBW). Landing and Westrop (1996) emphasized that the base of the massive, glauconitic (now metamorphosed to chlorite), coarse-grained sandstone 20.25–21.45 m above the base of section HBW (Figure 22) marked a stratigraphic break within a part of the succession dominated by fine-grained sandstones. This interpretation was based on the presence of phosphatic pebbles (with elements of the *Protolenus elegans* trilobite assemblage) and pebbles of Proterozoic vein quartz and rhyolite eroded from the Avalonian orogen at the base of the bed. In addition, deeply incised burrows, which are deeper than they are wide and were produced in overcompacted mud, occur at the base of the bed. Their presence is consistent with erosion prior to deposition of the coarse-grained sandstone (compare Landing and Brett, 1987).

Despite this evidence for a stratigraphic break, Landing and Westrop (1996) included this coarse-grained sandstone and 2.0 m of overlying fine-grained, laminated sandstone in the Somerset Street Member. The Somerset Street Member–Long Island Member contact was defined by those authors within the covered interval in the upper third of the Hanford Brook Formation section at Hanford Brook West (Figure 24).

In this report, the lower boundary of the Long Island Member is revised downward to correspond to the base of the massive, coarse-grained sandstone (20.25–21.45 m) at the representative section at Hanford Brook West (Figure 24). This revision involves a minor alteration in the upper boundary of the Somerset Street Member and the lower boundary of the Long Island Member at the only known locality where this boundary is exposed, and the original names of these members are retained (i.e., North American Commission on Stratigraphic Nomenclature, 1983, Article 19). The definitions of the Somerset Street and Long Island Members at their type sections, and the range of lithologies enumerated by Landing and Westrop (1996) and exhibited by these members at other localities, do not change. This contact also becomes more "natural" because it corresponds to a stratigraphic break in the type section of the Hanford Brook Formation.

The duration of this inter-member break is probably relatively short. This interpretation is based on the presence of *Protolenus elegans* assemblage trilobites through the underlying Somerset Street Member, in the reworked phosphate pebbles at the base of the massive sandstone, and as isolated specimens at the top of the massive sandstone.

The inter-regional significance within Avalon of the break between the Somerset Street and Long Island Members is presently uncertain. This break may compare with the depositional sequence 4A–4B break between the St. Mary's and Jigging Cove Members of the Brigus Formation in southeast Newfoundland (August 14, Jigging Cove, and August 16, Smith Point). Because depositional sequence boundaries are essentially synchronous across Avalon (Landing, 1996), and the sequence 4A–4B boundary corresponds to the contact of the *Callavia broeggeri* Zone and overlying *Myopsostrenua?* cf. *M.? sabulosa* Faunal Interval in southeastern Newfoundland, a temporal correlation of the Somerset Street–Long Island contact with the St. Mary's–Jigging Cove contact would require an important reevaluation of the biostratigraphic significance of *Protolenus*-bearing assemblages. Such a correlation would mean that the classic "*Protolenus* Zone" of the Hanford Brook Formation would not be high in the trilobite-bearing Lower Cambrian (e.g., Geyer and Palmer, 1995), but would be a faunal assemblage in sandstone facies that would correlate with the *Callavia broeggeri* Zone in siliciclastic mudstones and nodular limestones.

Manuels River Formation (Revised)

Manuels River Formation (revised).—The Manuels River Formation (Hutchinson, 1962) is revised to include an unconformity-bound, dark grey to black siliciclastic mudstone-dominated interval. The Manuels River Formation has thin-bedded sandstones, minor nodular (methanogenic) and bedded limestones, thin volcanic ashes, and local thick volcanic flows. The unit crops out from southeastern Newfoundland to southern New Brunswick, and likely as far south as eastern Massachusetts. In most areas of eastern Newfoundland, the base of the formation is defined by a volcanic ash-mantled unconformity surface, and the top of the formation is an unconformity with the uppermost Middle Cambrian Cavendish Formation (new) of the MacLean Brook Group (revised). In southeastern Newfoundland, the Manuels River Formation typically measures 20–30 m thick, with a maximum thickness of 43.5 m at Deep Cove in western St. Mary's Bay in the southern St. Mary's–east Trinity axis (Figure 2, locality DC; E. Landing and S.R. Westrop, unpub. data). The thickest sections of the Manuels River Formation apparently occur in Cape Breton Island (Figures 1, 20), where the formation has been mapped as the ca. 150 m-thick

"Trout Brook Formation" (Hutchinson, 1952; designation abandoned by Landing, 1996). However, the outcrop belt of the Manuels River Formation in Cape Breton Island is structurally complex, poorly exposed, and not measurable with any precision (E. Landing, unpublished data).

Although Hutchinson (1962) specified a black shale-dominated lithology of the Manuels River Formation, his original definition included Howell's (1925) beds 26–125 at Manuels River as the type section of the formation. As noted by Fletcher (1972), the number "26" was apparently a typographic error, because this bed falls within characteristic green siliciclastic mudstones of the underlying Chamberlain's Brook Formation. Howell's (1925) bed 36, by contrast, is a K-bentonite that serves as a good marker between the greenish shales and interbedded limestones with an *Eccaparadoxides eteminicus* Zone assemblage of the Fossil Brook Member (depositional sequence 7) of the uppermost Chamberlain's Brook, and overlying darker shales with a *Hydrocephalus hicksi* Zone fauna (August 12, Red Bridge Road). This basal K-bentonite at the Manuels River is regionally traceable (August 17, Fosters Point) and is a correlatable base of the Manuels River Formation.

As discussed above (August 12, Red Bridge Road and Manuels River), the Chamberlain's Brook–Manuels River contact is not conformable, as believed in earlier reports (Howell, 1925; Hutchinson, 1962; Fletcher, 1962). It shows differential erosion of up to 4 m on the underlying Chamberlain's Brook Formation and constitutes the depositional sequence 7–8 boundary in the Cambrian of Avalon (Landing, 1996; Figure 3).

A revision of the Manuels River Formation must also change the upper contact of the formation. Hutchinson (1962), Fletcher (1972), and Poulsen and Anderson (1975) brought a wide range of lithologies that overlie the typical black shales of the Manuels River Formation into this formation. These overlying units include: 1) Howell's (1925) bed 125, the thin conglomeratic sandstone immediately above the black shales at Manuels River (August 12, Figure 7) and Fosters Point (August 17, Figure 17); 2) an interval with a stromatolitic limestone and overlying sandstones at Highland Cove (August 13, Figure 10); and 3) 30 m of pillow basalts (Chapel Arm Member of McCartney, 1967). These higher lithologies were earlier included in the Manuels River Formation because limited fossil control showed that they were upper Middle Cambrian, as the Manuels River Formation.

As noted above at stops on August 12, 13, and 17, sandstones, a conglomeratic sandstone, a local stromatolitic limestone bed, and the Chapel Arm volcanics form a separate lithosome above the dark shales of the Manuels River Formation. These units are referable to the Cavendish Formation (new) at the base of the MacLean Brook Group (revised), and comprise the lowest part of depositional sequence 9 (uppermost Middle Cambrian–lower Upper Cambrian) of the Avalonian Cambrian (Landing, 1996). The Manuels River–Cavendish contact is demonstrably unconformable, and the uppermost part of the Manuels River Formation (i.e., the black mudstones of member 4 (to 6.2 m thick; August 13, Highland Cove, Figure 10) are only locally preserved because of post-Manuels River and pre-Cavendish erosion. This minor revision of the upper contact of the Manuels River Formation only deletes Howell's (1925) bed 125 (an 8–15 cm-thick yellow sandstone with shale and phosphate pebble clasts) from Hutchinson's (1962) definition of the type section.

Members 1–4 (new) of the Manuels River Formation.—Fletcher (1972) noted that four member-level units of the Manuels River Formation, defined by a four-fold vertical alternation of siliciclastic mudstone colors (dark grey to black and dark grey to black) are recognizable in the thickest section of the formation at Deep Cove in western St. Mary's Bay. We concur with this observation, and recognize these units across southeast Newfoundland. These informal members have been used above in the field trip guide of this volume, and a brief summary of them is provided below. The significance of these color alternations is that they appear to record alternations of weakly dysaerobic (dark grey) to more strongly dysaerobic (black) bottom environments during deposition of the Manuels River Formation.

Member 1 (new).—The lowest part of the Manuels River Formation is a dark olive green to medium-grey siliciclastic mudstone-dominated lithosome that tends to have very thin, locally wave-rippled, micaceous sandstone laminae, storm-winnowed trilobite hash lags, and up to four very thin K-bentonites. Phosphate granule lags are locally abundant and reflect the influence of storm events in deposition. Small, flattened, calcareous nodules develop on the sandstone laminae. Shallow burrows *(Planolites)*; hyolithids, many of which feature articulated conchs, opercula, and helens; and trilobites (*Hydrocephalus hicksi* Zone) are more abundant in member 1 than in the darker mudstones of member 2, and this suggests a more oxygenated sediment–water interface.

The base of member 1 is the K-bentonite-mantled unconformity at the base of the Manuels River Formation. Presence of this volcanic ash is locally important, because the locally developed olive-green color of this lower part of the Manuels River Formation is frequently similar to that of the siliciclastic mudstones of the underlying Fossil Brook Member of the Chamberlain's Brook Formation. Member 1 is capped by the bentonite at the base of member 2 of the Manuels River Formation.

Member 1 occurs at the base of the Manuels River Formation across southeastern Newfoundland. Characteristic thicknesses are on the order of 5.7–5.9 m (August 13, Manuels River; August 17, Fosters Point), and the thickest sections on the St. Mary's–east Trinity axis range from 8.6 m at Highland Cove (August 13) to 11.9 m at McLeod Point at the south end of Trinity Bay (E. Landing and S.R. Westrop, unpub. data).

Member 2 (new).—Member 2 consists of dark grey to black, pyritiferous, laminated, siliciclastic mudstones with sparse burrows. The member has large, relatively unfossiliferous, methanogenic, calcareous nodules in its lower several meters. The base of member 2 is a relatively thick (2–4 cm), geographically widespread K-bentonite, and the unit progressively darkens in color above the basal ash (August 12 Manuels River; August 17, Fosters Point). Trilobites and other benthic elements are rarer in member 2 than in member 1. The *Hydrocephalus hicksi* Zone assemblages of member 1 persist into member 2, but become sparser with the upward darkening of the shale. Member 2 ranges from 8—12 m in thickness (August 13, Highland Cove; August 17, Fosters Point).

Member 3 (new).—Member 3 represents a return to somewhat lighter grey siliciclastic mudstones. This color change takes place slightly below (ca. 0.5–1.1 m) a volcanic ash or bundle of ashes in the lower part of the member. The thin micareous sandstones, abundant shallow burrows, trilobite hash lenses, and thin calcareous nodules noted below in member 1 reappear in this similar lithofacies in member 3. The reappearance of a somewhat higher density of burrowing metazoans and probable higher bottom-water oxygenation in lower member 3 is associated with the appearance of a number of horizons with articulated *Paradoxides davidis* specimens in storm-lag beds. These articulated specimens were the subject of Bergstrom and Levi-Setti's (1978) study of *P. davidis*. At most localities east and west of the St. Mary's–east Trinity axis, member 3 composes a 6–8 m-thick top of the Manuels River Formation under the Cavendish Formation (e.g., August 17, Fosters Point).

Member 4 (new).—Member 4 of the Manuels River Formation is a very dark grey to black, siliciclastic mudstone- and siltstone-dominated facies with thin, normally graded sandstones and trilobite-hash lenses that is preserved below the unconformity with the Cavendish Formation only along the St. Mary's–east Trinity axis. 6.2 m and 8 m thicknesses of member 4 are present, respectively, at Highland Cove (August 13) in east Trinity Bay and at Deep Cove, western Trinity Bay (Figure 2, locality DC; E. Landing and S.R. Westrop, unpub. data). The middle part of member 4 features the reappearance of articulated trilobite assemblages with complete *Paradoxides davidis* specimens.

Important volcanism took place during member 4 deposition in the southern part of the St. Mary's–east Trinity axis. Up to 60 m of calcitized basaltic tuffs occur as a lens within member 4 at Hay Cove, western St. Mary's Bay (Fletcher, 1972). These tuffs thin to the south within a few kilometers to a 30 cm-thick calcitized ash bed in the middle of member 4 at Deep Cove (E. Landing and S.R. Westrop, unpub. data). Although

Fletcher (1972, p. 132) proposed that the Hay Cove volcanics are "essentially contemporaneous" with the Chapel Arm Member volcanics (McCartney, 1967) in southern Trinity Bay (discussed below), the Hay Cove volcanics are preserved at the top of depositional sequence 8, and the Chapel Arm volcanics are assigned to the Cavendish Formation (new) as an extrusive body in lowest depositional sequence 9.

CAVENDISH FORMATION (NEW)

Cavendish Formation (new).—The Cavendish Formation unconformably overlies the Manuels River Formation across southeastern Newfoundland and is a lenticular lithosome dominated by fine- to medium-grained, micaceous, and glauconitic sandstone, typically with a thin basal conglomerate, a local volcaniclastic sandstone and stromatolitic limestone (August 13, Highland Cove; Figure 10), and a lensing, thick (to 30 m) pillow basalt (Chapel Arm Member of McCartney, 1967). The top of the Cavendish Formation lies at a sharp contact with overlying black, pyritiferous, somewhat micaceous siliciclastic mudstones with sideritic nodules. The type section of the Cavendish Formation is at Highland Cove (Figure 10, August 13). Although this locality has been traditionally designated "Highland Cove" in the geological literature (e.g., Hutchinson, 1962; Poulsen and Anderson, 1975), the locality lacks any name on official maps. For this reason and because of the lack of an appropriate map name near this locality, the name of the formation is taken from Cavendish village, approximately 1.1 km to the southeast.

Thickness of the Cavendish Formation ranges from a maximum of 45 m in St. Mary's Bay (discussed below) to an 8–15 cm-thick featheredge of pyritiferous sandstone with black shale and phosphate clasts at Manuels River (August 12; Howell, 1925, bed 125) and Fosters Point (August 17). This conglomeratic sandstone thickens to 30–50 cm at Red Bridge Road, where it is overlain by 60 cm of thin bedded, weakly calcareous sandstones.

The 5.75 m-thick type section of the Cavendish Formation overlies the Manuels River Formation at Highland Cove, eastern Trinity Bay, and shows a basal stromatolitic limestone that overlies 5 cm of oxidized (green) shale at the top of the Manuels River Formation (August 13, Figure 10). Most of the Cavendish Formation at Highland Cove consists of the same type of fine-grained, burrowed (*Teichichnus* Seilacher), glauconitic, weakly calcareous sandstone that dominates the thin (1.1 m) reference section at Red Bridge Road. This thickening of the upper calcareous sandstone of the Cavendish Formation may be explained by location of the Highland Cove section in the more rapidly subsiding St. Mary's-east Trinity axis. An additional unit that appears at Highland Cove is a 15 cm-thick, dark purple, volcaniclastic sandstone in the middle of the formation. This sandstone reflects the eruption of pillow basalts that form the top of the Cavendish Formation at McLeod Point at the south end of Trinity Bay (Chapel Arm Member of McCartney, 1967).

In an unpublished dissertation, Fletcher (1972) recognized up to 45 m of greenish grey, thin-bedded alternations of fine-grained, calcareous, glauconitic sandstone and siltstones above the Manuels River Formation on the west side of St. Mary's Bay. This lithosome in the southern part of the St. Mary's–east Trinity axis is lithologically comparable and coeval with the dominant facies of the Cavendish Formation at its type section. Interestingly, Fletcher (1972) reported a 30 cm-thick "limestone" immediately underneath this sandstone–siltstone lithosome and included it as the top bed of the Manuels River Formation. This "limestone" proves to be a limestone clast conglomerate with paradoxidiid trilobite fragments in arenaceous, trilobite wackestone cobbles to 10x15 cm in diameter, and constitutes a conglomerate at the base of the sandstone-siltstone lithosome. Fletcher's (1972) name "Beckford Head Formation" for these post-Manuels River clastics is an unpublished synonym for the sandstone facies of the Cavendish Formation.

The Cavendish Formation is upper Middle Cambrian. As noted above (Landing and Westrop, this volume, August 13, "Highland Cove: biostratigraphy of sequence 8–9 unconformity"), a number of trilobites known from the *Paradoxides davidis* Zone of the upper Manuels River Formation appear in the Cavendish Formation. The lower Upper Cambrian agnostoid *Agnostus pisiformis* appears 10–18 m above the Cavendish Formation base, and this suggests that the formation does not range into the Upper Cambrian.

Comparison with Avalonian Britain.—The lithologies and timing of the Manuels River–Cavendish unconformity in southeast Newfoundland are recognizable in Avalonian Britain (Landing, 1996). On St. Tudwal's Peninsula, North Wales, the dark grey-black, fine-grained siliciclastics with thin volcanic ashes of the Nant-y-big formation are very similar to and coeval with (i.e., *Hydrocephalus hicksi-Paradoxides davidis* Zones) the Manuels River Formation (see Young et al., 1994). Similarly, the Abbey Shales of Warwickshire, England are coeval with and lithologically identical to the Manuels River Formation (Illing, 1916; Taylor and Rushton, 1971). In both of these British areas, micaceous sandstones and grey shales (Maentrog and Mancetter Formations, respectively) overlie the black facies, with a thin (up to 1 m) sandstone or calcareous clast conglomerate at the contact. Shergold and Brasier (1986) termed this locally stromatolitic bed at the base of the Maentwrog Formation the "*Dorypyge* Limestone", although the unit is primarily an amalgamated, condensed calcareous sandstone (Young et al., 1994). Paradoxidiid fragments occur in the matrix of the "*Dorypyge* Limestone" (E. Landing and S.R. Westrop, unpublished data), and Nicholas (1915, 1916) was apparently correct in assigning this bed to the uppermost Middle Cambrian, rather than to the lower Upper Cambrian (compare Young et al., 1994). Similarly, presence of uppermost Middle Cambrian trilobites of the *Lejopyge laevigata* Zone in the lowest Mancetter Formation (Illing, 1916; Taylor and Rushton, 1971) demonstrates that the calcareous pebble conglomerate at its base is uppermost Middle Cambrian.

These stratigraphic relationships allow recognition of a comparable late Middle Cambrian sequence stratigraphic history in Avalonian North America and Britain. In addition, they can be used to suggest that the lithologic similarities of sequence 8 and lower sequence 9 might be the basis for a unified stratigraphic nomenclature of these presently separated regions of Avalon. In this case, an Abbey Formation (Illing, 1916) would be senior synonym to the Manuels River (Hutchinson, 1962) and Nant-y-big (Young et al., 1994) Formations, and the Mancetter Formation (Taylor and Rushton, 1971) is senior synonym of the Maentwrog (Allen et al., 1981) and Cavendish (this report) Formations.

MacLean Brook Group (Revised)

Landing (1996) used Hutchinson's (1952) MacLean Brook Formation as the senior synonym for an uppermost Middle Cambrian–lower Upper Cambrian sandstone- and siltstone-dominated lithosome (i.e., depositional sequence 9) that unconformably overlies the Manuels River Formation in Avalonian North America. The proposal in this volume of a formation-level unit (Cavendish Formation [new]) above the Manuels River Formation requires a minor revision by a change in rank (see North American Commission on Stratigraphic Nomenclature, 1983) of the MacLean Brook "Formation" to an expanded "MacLean Brook Group".

Acknowledgments

This synthesis was done under National Science Foundation grant EAR94-15773 (to E. Landing) and Natural Sciences and Engineering Research Council grant 41197 (to S.R. Westrop).

References

Allen, P.M., A.A. Jackson, and A.W.A. Rushton. 1981. The stratigraphy of the Mawddach Group in the Cambrian succession of North Wales. Proceedings of the Yorkshire Geological Society, 43:295–329.

Bergstrom, J., and R. Levi-Setti. 1978. Phenotypic variation in the Middle Cambrian trilobite *Paradoxides davidis* Salter at Manuels, SE Newfoundland. Geologica et Palaeontologica, 12:1–40.

Brasier, M.D. 1989. Sections in England and their correlation, p. 82–104. *In* J.W. Cowie and M.D. Brasier (eds.), The Precambrian–Cambrian Boundary. Oxford Monographs on Geology and Geophysics, 12, 213 p.

Fletcher, T.P. 1972. Geology and Lower to Middle Cambrian trilobite faunas of the southwest Avalon, Newfoundland. Unpub. Ph.D. dissertation, University of Cambridge, 520 p.

Geyer, G., and A.R. Palmer. 1995. Neltneriidae and Holmiidae (Trilobita) from Morocco and the problem of Early Cambrian intercontinental correlation. Journal of Paleontology, 69:459–474.

Howell, B.F. 1925. The faunas of the Cambrian *Paradoxides* beds at Manuels, Newfoundland. Bulletins of American Paleontology, 11(3), 140 p.

———, and C.J. Stubblefield. 1950. A revision of the fauna of the north Welsh *Conocoryphe viola* beds implying a Lower Cambrian age. Geological Magazine, 87:1–16.

Hutchinson, R.D. 1952. The stratigraphy and trilobite faunas of the Cambrian sedimentary rocks of Cape Breton Island, Nova Scotia. Geological Survey of Canada Memoir 263, 124 p.

———. 1962. Cambrian stratigraphy and trilobite faunas in southeast Newfoundland. Geological Survey of Canada Bulletin 88, 156 p.

Illing, V.C. 1916. The paradoxidian fauna of a part of the Stockingford Shales. Quarterly Journal of the Geological Society of London, 71:386–448.

Landing, E. 1988. Lower Cambrian of eastern Massachusetts: stratigraphy and small shelly fossils. Journal of Paleontology, 62:661–695.

———. 1992. Lower Cambrian of southeastern Newfoundland: epeirogeny and Lazarus faunas, lithofacies-biofacies linkages, and the myth of a global chronostratigraphy, p. 283–309. *In* J.H. Lipps and P.W. Signor (eds.), Origin and Early Evolution of the Metazoa. Plenum Press, New York.

———. 1995. Upper Placentian–Branchian Series of mainland Nova Scotia (middle–upper Lower Cambrian): faunas, paleoenvironments, and stratigraphic revision. Journal of Paleontology, 69:475–495.

———. 1996. Avalon: insular continent by the latest Precambrian, p. 29–63. *In* R.D. Nance and M.D. Thompson (eds.), Avalonian and Related Peri-Gondwanan Terranes of the Circum-North Atlantic. Geological Society of America Special Paper 304.

———, and A.P. Benus. 1988. Stratigraphy of the Bonavista Group, southeastern Newfoundland: growth faults and the distribution of the sub-trilobitic Lower Cambrian, p. 59–71. *In* E. Landing, G.M. Narbonne, and P. Myrow (eds.), Trace Fossils, Small Shelly Fossils, and the Precambrian–Cambrian Boundary. New York State Museum Bulletin 463.

———, and C.E. Brett. 1987. Trace fossils and regional significance of a Middle Devonian (Givetian) disconformity in southeastern Ontario. Journal of Paleontology, 61:205–230.

Lapworth, C. 1898. A sketch of the geology of the Birmingham district. Proceedings of the Geological Association of London, 15:313–416.

McCartney, W.D. 1967. Whitbourne Map-Area, Newfoundland. Geological Survey of Canada Memoir 341, 135 p.

Morris, T.O., and W.G. Fernsides. 1926. The stratigraphy and structure of the Cambrian slate belt near Nantlle (Carnarvonshire). Quarterly Journal of the Geological Society of London, 82:250–303.

Myrow, P.M., and E. Landing. 1993. Mixed siliciclastic-carbonate deposition in an Early Cambrian oxygen-stratified basin, Chapel Island Formation, southeastern Newfoundland. Journal of Sedimentary Petrology, 62:455–473.

Nicholas, T.C. 1915. The geology of the St. Tudwal's Peninsula (Carnarvonshire). Quarterly Journal of the Geological Society ofLondon, 71:83–143.

———. 1916. Notes on the trilobite faunas of the Middle Cambrian of the St. Tudwal's Peninsula (Carnarvonshire). Quarterly Journal of the Geological Society of London, 71:451–472.

North American Commission on Stratigraphic Nomenclature. 1983. North American Stratigraphic Code. American Association of Petroleum Geologists Bulletin, 67:841–875.

Poulsen, V., and M.M. Anderson. 1975. The Middle–Upper Cambrian transition in southeastern Newfoundland, Canada. Canadian Journal of Earth Sciences, 12:2065–2079.

Rushton, A.W.A. 1966. The Cambrian trilobites of the Purley Shales of Warwickshire. Palaeontological Society Monograph, 511, 36 p.

Taylor, and A.W.A. Rushton. 1971. The pre-Westphalian geology of the Warwickshire coalfield with a description of three boreholes in the Meredale area. Bulletin of the Geological Survey of Great Britain, 35, 150 p. (issued in 1972).

Young, T., F. Martin, W.T. Dean, and A.W.A. Rushton. 1994. Cambrian stratigraphy of St. Tudwal's Peninsula, Gwynedd, northwest Wales. Geological Magazine, 131:335–360.

Abstracts of Oral Presentations

Middle Cambrian Reefs and Hardgrounds in Iran

Hamdi, Bahaeddin, Geological Survey of Iran, P.O. Box 131851-1494, Tehran, Iran; ZHURAVLEV, Paleontological Institute, Russian Academy of Sciences, Profsoyuznaya ul. 123, Moscow 117647, Russia; and KRUSE, Pierre D., Northern Territory Geological Survey, P.O. Box 2901, Darwin, NT 0801, Australia

Thrombolites and sponge (lithistid) bafflestones compose reefs in the Mila Formation of the Elburz Mountains, Iran. Hardgrounds that developed on shell-hash limestones served as the substrate for reef initiation. Calcite cements and fossil fragments characterize these Middle Cambrian substrate limestones, although aragonite marine cements were common in the Early Cambrian. Articulate brachiopods and eocrinoid sclerites, with low-Mg-calcite composition, were the principal allochems in these substrate limestones. Thrombolite-lithistid reefs that first appeared in Iran in the late Middle Cambrian provided a blueprint for Late Cambrian–Early Ordovician reefs.

Structure, ecological composition, and even mineralogy of reefs changed drastically during the Cambrian. Early Cambrian reefs were dominantly composed of calcimicrobian dendrolites (e.g., microframeworks formed by calcified microbes) and sponges (mainly archaeocyathan) that formed mudmounds and bafflestones and, rarely, boundstones. Late Middle Cambrian and Late Cambrian reefs were constructed almost exclusively by thrombolites and stromatolites, which also dominated Early Ordovician buildups. The thrombolites commonly have a metazoan component in their composition. The replacement of Early Cambrian-type reefs by a thrombolite-stromatolite consortium featured: 1) the complete extinction of Early Cambrian reef-building metazoans; 2) the development of relatively large and abundant benthic animals with low-Mg-calcite sclerites that were cemented by calcite into hardgrounds; and 3) the probable replacement of aragonite-precipitating conditions into aragonite-inhibiting.

Cambrian Potpourri of Aussie Brachiopods and Iranian Sponges

Kruse, Pierre D., Northern Territory Geological Survey, P.O. Box 2901, Darwin, NT 0812, Australia

Because trilobites common to both regions are unknown, there has been some lingering uncertainty in correlating the mainly Lower Cambrian succession of South Australia with the mainly Middle Cambrian (in Australian terms) and younger succession in central and northern Australia. For example, a recent correlation regards the *Archaeocyathus abacus* beds of the Wirrealpa Limestone (South Australia) as older than the Ordian–lower Templetonian *Xystridura templetonensis–Redlichia chinensis* interval that is widespread through central and northern Australia.

Recent studies of brachiopods from the latter interval in northern Australia affirm, in fact, that these two intervals are best considered coeval and correlate with the Toyonian and perhaps Amgan stages of Siberia. The acrotretid *Hadrotreta djagoran* and acrothelid *Eothele napuru*, first described from the Daly Basin in northern Australia, have now been recognized from the Wiso and Georgina Basins of the same region, from the Wirrealpa Limestone of South Australia, and even from an allochthonous boulder on the Antarctic Peninsula.

In the late Middle Cambrian of northern Iran, the Mila Formation preserves the oldest known spiculate sponge reefs. These are bioherms developed in a high-energy milieau. They incorporate an anthaspidellid sponge described as *Rankenella mors*, microbial encrustations, and a modicum of calcimicrobes and fibrous and equant cement. These bioherms provide a rare glimpse of metazoan bioconstruction in the generally metazoan reef-depauperate Middle and Upper Cambrian. By comparison with Early Cambrian bioherms, the sponges and microbial encrustations played analogous bioconstructional roles to those of the Early Cambrian archaeocyathans and calcimocrobes, and represent frame building and encrusting/binding roles, respectively.

Global Significance of the Late Cambrian Steptoean Positive Carbon Isotope Excursion (SPICE)

Runnegar, Bruce, and Saltzman, Matthew R., Department of Earth and Space Sciences and Institute of Geophysics and Planetary Physics, University of California, Los Angeles, California 90095

A large positive excursion (+4°/$_{oo}$) in the $^{13}C/^{12}C$ ratios of Late Cambrian carbonates has been discovered in Australia (Georgina Basin), People's Republic of China (Hunan), Kazakhstan, and North America (Great Basin and upper Mississippi River valley). The event begins at a global mass extinction of trilobites that defines the base of the Steptoean Stage in Laurentia, the Idamean in Australia, and the Changshanian in China. The timing and duration of this Steptoean Positive Carbon Isotope Excursion (SPICE) are constrained by the average of several SHRIMP U-Pb zircon ages on Boomerangian volcanics in Tasmania (503 $\pm$ 3.5 Ma; see Perkins and Walshe, 1993), which predate the beginning of the SPICE event by one Australian stage (Idamean), and a precise, conventional (IDTIMS) U-Pb zircon age from the upper Upper Cambrian of Wales (lower *Peltura scarabaeoides* Zone, 491 $\pm$ 1 Ma; Davidek et al., In press), which post-dates the SPICE excursion by four trilobite zones. By interpolation, it seems as if the SPICE excursion lasted for some 3 Ma and began about 500 Ma ago. Numerical modelling with Broeker and Peng's one-dimensional, two-box model, as developed by Kump (1991), also suggests a 1–2 Ma duration for the SPICE event. Although the ultimate cause of this major perturbation of the Cambrian carbon cycle remains unclear, there are clear parallels and also differences with younger extinctions coupled with excursions (i.e., the Frasnian–Fammenian, Cenomanian–Turonian, and K–T events). The survival of the primary (heavy) carbon isotope signature, even in dolomitized or highly altered rocks, such as those that host Mississippi Valley-type (MVT) ore deposits, is a clear indication that the SPICE excursion will be a valuable tool for precise regional and intercratonic correlation of Upper Cambrian strata and events, even in deformed terranes. As it began at or near the base of the *Glyptagnostus reticulatus* Zone, the SPICE excursion may eventually be used globally to locate one of the major subdivisions of the Cambrian System in unfossiliferous or sparsely fossiliferous carbonate or carbonaceous sequences.

The Leonian Stage (Lower–Middle Cambrian): A Chronostratigraphic Unit for the Mediterranean Subprovince

Sdzuy, Klaus, Institute fur Palaontologie der Universitat Wuerzburg, Pleicherwall 1, 97070 Wuerzburg, Germany; LINAN, Eladio, Departamento de Geologia, Universidad de Zaragoza, 50009 Zaragoza, Spain; and GOZALO, Rodolfo, Departamento de Geologia, Universidad de Valencia, C/ Dr. Moliner 50, 46100 Burjasot, Spain

The Leonian Stage is a chronostratigraphic unit formally defined in the Iberian Peninsula as the lowest stage of the Middle Cambrian series (Figure 29). It has been correlated with trilobites into the *Paradoxides oelandicus* Stage of Baltica and the Amgan Stage of Siberia (in part). It is presented herein as a regional stage for the Mediterranean Subprovince because it contains biostratigraphic markers recognizable in sublittoral and open marine facies in Spain, Portugal, Italy, France, Morocco, Germany, and Turkey.

The lower boundary of the Leonian Stage is defined at the lowest appearance datum of *Eoparadoxides mureroensis*. This datum is one of the most widespread for the Middle Cambrian because it is recorded in the Cantabrian Mountains and Iberian Chains (Spain), Sardinia (Italy), the High Atlas and Anti-Atlas (Morocco), the Taurus Mountains (Turkey), Tuva (Siberia), and possibly eastern Newfoundland (Canada). The upper boundary of the stage is defined at the lowest occurrence of *Badulesia tenera*, as recognized in eastern Massachusetts (USA); southern New Brunswick and eastern Newfoundland (Canada); the High Atlas (Morocco); the Sierra Moreno, Cantabrian Mountains, and Iberian Chains (Spain); Doberlug (Germany); and Amanos Mountains (Turkey).

Three Leonian trilobite zones are recognized: *Eoparadoxides mureroensis*, *Paradoxides (Eccaparadoxides) sdzuyi*, and *P. (E.) asturianus* Zones. These zones are characterized by a sequence of bioevents that occur within this stage in Spain and other countries.

The middle Leonian regression that took place at the end of the *Paradoxides (Eccaparadoxides) sdzuyi* Chron allows integration of the Leonian Stage into the record of Cambrian transgressive–regressive events. The earliest Leonian transgressive episode corresponds to the *Eoparadoxides mureroensis* and most of the *P. (E.) sdzuyi* zones. Based on the known Iberian fossil record, this first interval featured a rapid increase in trilobite diversity that led to the appearance of at least 24 new genera and 35 new species of polymeroids and miomeroids and represented a major interval of trilobite diversification in the Cambrian. The Leonian regressive interval corresponds to an important fall in diversity with the disappearance of sixteen trilobite species. The second Leonian transgressive interval featured the appearance of three new genera and 21 new species. These data indicate that an event-stratigraphic analysis of the Leonian provides another tool to characterize and interpret the trilobite evolutionary record, and can be used in inter-regional correlation to complement available biostratigraphic data.

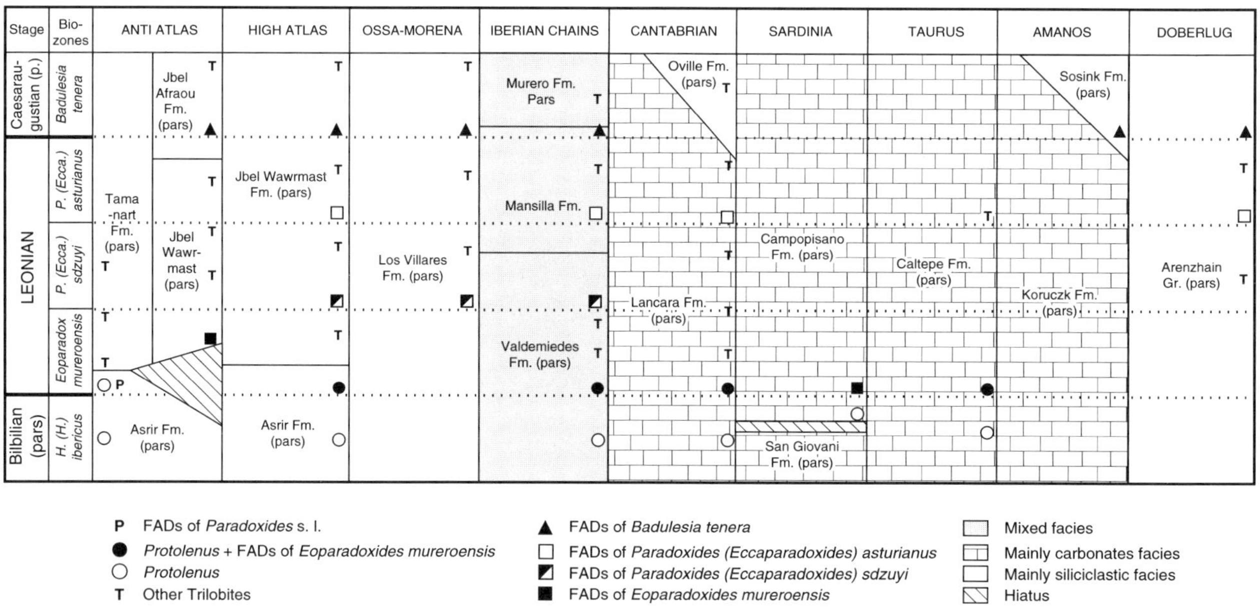

Figure 29. Correlations of Leonian Stage (lower Middle Cambrian) and sub- and superjacent strata in the Mediterranean Subprovince.

THE END OF MIDDLE CAMBRIAN TIME

Shergold, John H., Institute des Sciences de l'Evolution, Laboratoire de Paleontologie, Universite de Montpelier II, F34095 Montpelier Cedex 05, France

If international agreement exists and a Middle Cambrian series is retained, at which biostratigraphical datum should it end? Whatever choice is made defines the beginning of Late Cambrian time. The horizon selected should also coincide with the base of an appropriately defined stage. Of the three choices available, only one meets the conditions required by the International Commission on Stratigraphy. The traditional base of the Upper Cambrian is at the base of the *Agnostus pisiformis* Zone, which is confined to the Avalonian and Scandian faunal provinces. Unfortunately, it has an impoverished fauna in Scandinavia, which limits its division into finer biostratigraphic units, and its base is not well defined there, although it has been widely correlated. It essentially marks the base of the Olenid Series in Scandinavia and the Merioneth Series in Avalon, but geographically defined stages are unavailable in these regions. It, however, also marks the base of the Tuorian Stage in Siberia (base of *Agnostus pisiformis-Proagnostus fecundus* Zone). This horizon correlates into the Dresbachian Stage of Laurentia, the Kushan Stage of China, the Ayusokkanian Stage of Kazakhstan, and the Mindyallan Stage of Australia.

The Tuorian Stage also embraces the *Glyptagnostus stolidotus* and *Glyptagnostus reticulatus* Zones, both of which have been regarded as defining the Middle–Late Cambrian boundary. The *Glyptagnostus stolidotus* Zone is geographically widespread, and has a large fauna that enables interprovincial correlation. This species represents a First Appearance Datum (FAD) of *Glyptagnostus*, and its lowest occurrence has been regarded as marking the base of the Upper Cambrian, particularly in south-central China (northern Sichuan, southern Shaanxi, Henan, western Hunan, eastern Guizhou), but nowhere does its FAD coincide with a correlatable stage boundary. *Glyptagnostus reticulatus* also has a cosmopolitan occurrence, and its FAD has been taken to define the base of the Upper Cambrian in Kazakhstan. It occurs at the bases of the Olenus Series in Baltoscandia and Avalon, Changshanian Stage in China, Idamean Stage of Australia, and the Steptoean Stage in Laurentia. In many parts of the world, there is a faunal turnover between the *Glyptagnostus stolidotus* and *Glyptagnostus reticulatus* Zones. The *Glyptagnostus reticulatus* Zone is recognized by all workers as indisputably Upper Cambrian, and must rank as a serious contender as the upper bracket of the Middle Cambrian.

NEOPROTEROZOIC–ORDOVICIAN BIOTIC DIVERSITY

Zhuravlev, Andrey Yu., Paleontological Institute, Russian Academy of Sciences, Profsoyuznaya ul. 123, Moscow, 117647, Russia

Data on 4,122 metazoan genera, 31 calcimicrobial genera, and 470 acritarch species were plotted for the Nemakit-Daldynian–early Tremadoc at the zone level. Generally congruent plots of the diversity of metazoan genera, acritarch species, calcified cyanobacteria, and ichnofossils characterize Nemakit-Daldynian–early Botomian diversification, the middle Botomian crisis that led to further late Botomian–Toyonian diversity decreases, and the low diversity stabilization of the Middle–Late Cambrian. Coordinated fluctuations in numbers of monotypic taxa, geographic distribution, and longevity indices among different groups are present. Decreases in average longevity and geographic distribution indices took place before extinction events, and this shows that ecologic specialists, which are characterized by short longevity and high endemism, vanished during extinction events. At the same time, the average monotypy index decreases to show the appearence of more numerous large families that contain a number of genera, a feature that increases niche overlap and competition. Thus, the existing structure of ecosystems, which may be expressed by the relative number of specialists and degree of competition, was responsible for biotic stability. It is believed that extrinsic factors can amplify crises, but cannot initiate them.

The support of the Russian Foundation for Basic Research project 96-05-64224 is acknowledged.

DIVERSITY OF CAMBRIAN CALCIFIED ALGAE AND CYANOBACTERIA(?)

Zhuravlev, A.Yu., Paleontological Institute, Russian Academy of Sciences, Profsoyuznaya ul. 123, Moscow 117647, Russia

The type and topotype material of 133 Cambrian genera regarded as calcified algae or cyanobacteria has been examined. Only three of these genera (Middle Cambrian *Amgaella* and Late Cambrian *Seletonella* and *Mejerella*) can be assigned to true algae, but are of uncertain systematic position.

Twenty-six genera represent a group of calcified microfossils of uncertain affinities, which are commonly termed "calcimicrobes". Their organization is very simple by comparison with true algae, but their distinct microstructure does not allow assignment to any procaryotic group. These fossils have been regarded as Calcibionta or Cyanophycea, but may be better considered as renalcids. Cambrian renalcids include:

1) botryoidal forms (*Renalcis, Angusticellularia* [commonly reported as *Angulocellularia*], *Chabakovia, Gemma, Tarthinia*, and a new genus)
2) dendritic forms (*Epiphyton, Gordonophyton, Kordephyton, Korilophyton, Parepiphyton, Tubomorphophyton*)
3) conical–hemispherical forms (*Acanthina, Amgaina, Sajania*)
4) tuberous forms (*Bija, Bajanophyton, Botomaella, Nuratella*)
5) coiled filaments (*Obruchevella*)
6) straight filaments (*Girvanella, Razumovskia*)
7) bundles (*Batinevia, Subtifloria*)
8) tubular bush-like forms (*Proaulopora*)
9) chambered forms (*Wetherelella*).

One hundred other genera that have been described as calcified algae and cyanobacteria from the Cambrian are reinterpreted as: metazoan skeletons (e.g., *Cambroporella, Edelsteinia, Kadvoya* [commonly cited as the earliest red algae]), nonbiologic precipitates (e.g., *Comansita, Mawsonella, Nuia, Zaganolomia*), noncalcified microbian structures (e.g., *Filiphycus, Protuberantia, Uranovia*), trace fossils (e.g., *Aulophycus, Ketemella*), and all junior synonyms of the forms listed above.

The support of the Russian Foundation for Basic Research, Project 96-05-64224 is gratefully acknowledged.